Fungi

by René Fester Kratz, PhD

for
dummies®
A Wiley Brand

Fungi For Dummies®

Published by: **John Wiley & Sons, Inc.**, 111 River Street, Hoboken, NJ 07030-5774, www.wiley.com

Published simultaneously in Canada

The manufacturer's authorized representative according to the EU General Product Safety Regulation is Wiley-VCH GmbH, Boschstr. 12, 69469 Weinheim, Germany, e-mail: Product_Safety@wiley.com.

For general information on our other products and services, please contact our Customer Care Department within the U.S. at 877-762-2974, outside the U.S. at 317-572-3993, or fax 317-572-4002. For technical support, please visit https://hub.wiley.com/community/support/dummies.

Wiley publishes in a variety of print and electronic formats and by print-on-demand. Some material included with standard print versions of this book may not be included in e-books or in print-on-demand. If this book refers to media that is not included in the version you purchased, you may download this material at http://booksupport.wiley.com. For more information about Wiley products, visit www.wiley.com.

Library of Congress Control Number: 2025946196

ISBN 978-1-394-34483-3 (pbk); ISBN 978-1-394-34485-7 (ebk); ISBN 978-1-394-34484-0 (ebk)

Printed and bound by CPI Group (UK) Ltd, Croydon, CR0 4YY

C9781394344833_100925

Table of Contents

Introduction

Fungi may be the most mysterious and powerful organisms on the planet. Most of the time, they grow almost invisibly through the soil below your feet, decomposing plant matter. When environmental conditions are right, some of them make their presence known, exploding in a sudden profusion of colorful mushrooms. Many people around the world have appreciated the visible fungi for thousands of years, benefitting from their excellent nutritional qualities and powerful chemicals. Other people are newer members of the fungal fan club, attracted by new flavors, a desire to live more sustainably, or even just a longing to probe the secrets behind the mushroom. This book is about the ways fungi make all life possible and how they can make it better.

About This Book

Fungi For Dummies is an introduction to the world of fungi and their role in ecosystems. A special emphasis is placed on their importance to people. My goal is to provide the fundamentals of fungal structure and life history that you could apply to either the academic study of fungi or a personal interest in foraging or growing mushrooms. Depending on your personal background, this book may contain surprises about the importance of fungi to agriculture, medicine, or spirituality.

The table of contents highlights the modular organization of the book, which makes it easier for you to find the sections most useful to you. I present some fungal fundamentals in the first part of the book, then dive into some more technical details in the second part. The third part showcases some of the ways people use fungi, and the last part hits some of the highlights of why fungi are so cool.

Foolish Assumptions

As I wrote this book, I tried to imagine who you are and what type of information you are looking for. Here's what I came up with:

>> You may be someone who became interested in fungi because of their beauty and wants to know more about them. The chapters at the beginning and end of the book will fill you in on the fundamentals and the ways fungi are relevant to your life.

>> You may be someone who loves the flavors of fungi and wants to learn more about foraging or growing your own. There's a chapter on fungal structures to get you started on learning to identify fungi and a chapter that introduces you to growing fungi on your own. I've also provided suggestions for resources you can turn to when you're ready to take the next step.

>> You may be an experienced forager who would like to learn more about the scientific details of how fungi grow. If you have never had formal academic training on fungi and would like to know what you missed, the middle of the book is for you.

>> You may be a college biology major studying botany as part of your year-long freshman series or a student in an introductory mycology course. The beginning and middle of the book will fill you in on the fungal structures and life cycles that you'll need to know for class.

Icons Used in This Book

Throughout this book, icons in the margins highlight certain types of valuable information that call out for your attention. Here are the icons you'll encounter and a brief description of each.

TIP

The Tip icon lets you know what you need to do to get to the heart of the matter at hand. These icons mark information that helps you remember the facts being discussed or suggest a way to help you commit them to memory.

REMEMBER

Remember icons mark the information that's especially important to know. To siphon off the most important information in each chapter, just skim through these icons.

TECHNICAL STUFF

The Technical Stuff icon marks information of a highly technical nature that you can normally skip over.

WARNING

The Warning icon tells you to watch out! It marks important information that may save you headaches. I'll use this icon to let you know which old names for fungal groups are no longer being used, and which subjects are still under investigation.

Beyond the Book

In addition to the abundance of information and guidance related to fungi that I provide in this book, you get access to even more help and information online at Dummies.com. Check out this book's online Cheat Sheet, which provides a few concepts to get you started on your journey into the world of fungi and their role in Earth's ecosystems. Just go to www.dummies.com and search for "Fungi For Dummies Cheat Sheet."

Where to Go from Here

Like all *For Dummies* books, each chapter in *Fungi For Dummies* is self-contained, so you can pick it up whenever you need it and jump into the topic you are working on. You can start reading this book in whatever part interests you the most. If you want to know about the basic structures of fungi, how they grow, and their importance to ecosystems, you can start at the beginning. If you're more interested in how people use fungi, you can begin at the end. And if you're a student in a biology class, or just someone who wants to get into the finer details of different types of fungi, you can jump to the middle. Wherever you are, I'll give you tips about other places in the book that connect to what you're reading.

I hope you enjoy your journey into the world of fungi and find them as fascinating and beautiful as I do!

1

Getting Started with Fungi

Explore the diversity of Kingdom Fungi, from the mushrooms you see in the forest, to the yeast that's used to make bread and wine, to the fuzzy green mold that grows on old oranges, and many more you may not even notice.

Discover the role fungi play in nature and the important role they play in decomposition.

Take a look at how fungi partner with other organisms to help them grow.

Examine fungi at a cellular level and consider the features that make them unique among living things.

Explore how fungi reproduce and change their DNA over time.

Chapter **1**

Finding the Fungus Among Us

Kingdom Fungi includes everything from the mushrooms you see in the forest or the grocery store, to the yeast that's used to make bread and wine, to the fuzzy green mold that grows on old oranges, and many more you may not even notice. Fungi can be mysterious, suddenly appearing as a "fairy circle" in the forest. Some fungi are a little spooky, like the ones that turn ants into zombies or those that glow in the dark. Fungi can be beautiful too, like the fly agaric with its bright red cap and white spots (*Amanita muscaria*) or the veiled lady (*Phallus indusiatus*) shown on the cover of this book.

These weird and wonderful fungi lead secret lives beneath the surface of the soil or within the bodies of other organisms. They are essential to the cycle of life on Earth, and many can directly harm or benefit human health. This chapter introduces you to fungi and gives you a peek into what is happening below the surface of the fungal structures you can see.

REMEMBER

Mycology is the study of fungi, including their structure, role in nature, life cycles, and chemistry (*myco*=fungus, *logy*=study of).

Appreciating the Power of Fungi in Nature

Organisms interact with each other and their environment in complicated webs called *ecosystems*. For example, you're probably familiar with the way organisms interact in food chains. A *food chain* represents the movement of energy and nutrients through a series of organisms. For example, a mouse eats seeds, and then a snake eats the mouse.

When one organism eats another, it takes in the large molecules like carbohydrates, proteins, and fats that form the bodies of living things. It digests these large molecules into smaller components, taking the energy and nutrients it needs for its own growth. The molecules that make up one living thing get broken down and reformed into the molecules of another. If you eat a slice of pizza, you will essentially recycle some of the pizza molecules to build or repair your own body. (Yes, in some ways, you are what you eat.)

While all organisms pass nutrients through food chains, fungi are nature's ultimate recyclers. They make enzymes that can digest the really tough stuff like wood that few other organisms can touch (for details on how fungi break down wood, see Chapter 2). And they're not squeamish. They don't need their food to be alive or even freshly killed. In fact, they're specialists in digesting the dead.

REMEMBER

Decomposers digest dead organisms, breaking down their large molecules into smaller components. In the process, they release nutrients like minerals and carbon dioxide (CO_2) back into the environment where they can be picked up and reused by other living things (see Chapter 2 for more details on how carbon moves through ecosystems).

The molecules that make up living things are a type of matter. *Matter* is anything that has mass (can be weighed) and takes up space. It's the stuff that makes up everything around you, from the metal atoms in your cell phone to the cellulose that forms the paper in this book. Except for the occasional meteorite that hits the surface of the planet, no new matter enters our environment. (Light from the sun is energy, not matter.) All of the matter that makes up our planet and living things must be recycled if life is to continue.

Without decomposers like fungi and bacteria, there wouldn't be enough matter available for the growth of new life.

Exploring How Fungi Form Partnerships with Other Organisms

Many fungi form intimate relationships that facilitate the growth of other organisms.

Mycorrhizal fungi attach to plant roots and grow outward through the soil in vast networks. In exchange for food, the mycorrhizae increase the access of plants to water and minerals (Chapter 3). Experiments suggest mycorrhizae can even enable communication between plants over their mycelial network. Almost all plants, including those we grow for food, have mycorrhizal partners that support their growth.

Fungi also form partnerships with other photosynthetic organisms, like algae and blue-green bacteria, becoming lichens that grow on the surfaces of rocks, plants, and soil.

Lichens are plant-like growths that form from the symbiosis of at least two organisms, a fungus and a photosynthetic partner (Chapter 3).

The fungal mycelium provides a protective covering for its partner, storing water and blocking intense light. In return, the alga or bacterium shares some of the food it makes through photosynthesis.

Taking a Close Look at Fungal Structures

The parts of fungi that you can see are just a tiny fraction of the entire organism. When you see a mushroom, for example, you may think you're looking at a single individual, but what you're actually seeing is the reproductive structure of a massive organism that's growing beneath the mushroom.

Fungi grow as barely visible threads called *hyphae* that weave through soil and decaying organisms (Chapter 4).

Hyphae are long chains of fungal cells that grow and weave themselves throughout their environment, forming mats called *mycelia*. Sometimes, fungi shape their mycelia into large reproductive structures like mushrooms, bracket fungi, or puffballs. Examining the details of structures like these can help you identify wild fungi.

The reproductive structures of fungi produce *spores,* which are special cells that protect the genetic information of a new individual (Chapter 4). Fungi can clone themselves by producing spores asexually, or they can produce spores after a sexual process that combines genetic information from two individuals. The spores found among the gills on the bottom of a mushroom or in the pores of a bracket fungus are examples of sexually produced spores.

Unlocking the History of Fungi

Fungi have an interesting life cycle, which makes them useful organisms for the study of genetics (Chapter 6). The cells that form fungal hyphae are *haploid,* meaning they only have one copy of each gene. This makes it easier to study the effects of mutations on genes in fungi than it is in other organisms, like plants and animals, which typically have two copies of each gene.

Scientists think the first fungi lived in the water, then made their way onto land around the same time as plants, although they're still debating the exact order of events (Chapter 7). Scientists haven't studied many fossils of fungi, but some fossils that resemble modern aquatic chytrids (Chapter 8) can be dated to the late Precambrian period (650 to 543 million years ago). Fossils from the Devonian period (417 to 354 million years ago) show land plants and fungi forming symbiotic associations similar to today's mycorrhizal associations. In fact, some scientists believe that plants wouldn't have been able to move from the water to the land without the help of fungal partners.

Discovering the Many Groups of Fungi

The ability to read and compare the DNA of organisms caused many changes to the way scientists see the relationships between life on Earth. Prior to the development of DNA science, scientists classified organisms based on their physical characteristics, how they got their nutrition, and how they reproduced. These methods worked well for many larger species, but microscopic organisms with few visible characteristics were harder to sort out.

As a result, scientists have made many recent changes to the organization of the fungi. Some organisms that we thought were closely related to fungi, because of the way they look and grow, were kicked out of the kingdom. Relationships within the kingdom changed, and some fungi were moved from one group to another. And some fungi that never did scientists the courtesy of showing their sexual stages so that they could be sorted were finally put in their place on the basis of their DNA. Finally, the vast majority of fungi that grow in nature haven't been identified at all, so more changes to our understanding of the relationships of fungi to each other and other organisms will no doubt occur in the future.

Whether they are true fungi or not, these groups of organisms are commonly included in the science of mycology:

>> **Stramenopiles** include the water molds (oomycetes) and slime nets (labyrinthulomycetes). Both of these groups were originally included in Kingdom Fungi because of their fungal-like lifestyles.

>> **Amoebozoa** are the slime molds, amoebae that can often be found along with fungi growing on rotten logs. Also like fungi, they reproduce by spores and are important decomposers.

>> **Opisthosporidia** are intracellular parasites that don't have much surface resemblance to fungi, but whose DNA reveals a very close relationship (Chapter 8).

>> **Chytridiomycota,** also known as chytrids, are swimming single-celled organisms that decompose organic matter and have chitin in their cell walls. These characteristics, along with their DNA, put them in the fungal kingdom (Chapter 8).

>> The **zygomycetous fungi** include many of the familiar molds you've seen growing on your cheese, bread, and spoiled fruit (Chapter 9).

>> The **Ascomycota** is important to human food production because this group contains yeast, morels, and truffles, as well as some plant pathogens known as mildews (Chapter 10).

>> The **Basidiomycota** is the home for most of the familiar mushrooms as well as plant pathogens known as rusts and smuts (Chapter 11).

Making Connections Between Fungi and People

In addition to their importance to the health of the planet, fungi have many direct impacts on the lives of people:

>> **Food:** Fungi are not only delicious, but they're also packed with beneficial nutrients such as dietary fiber (Chapter 12). Plus, research shows they have a positive impact on your brain and immune system and contain compounds with many potential health benefits (Chapter 14).

>> **Spiritual uses:** Some cultures past and present use psychedelic mushrooms in their spiritual practices (Chapter 13). Doctors today are exploring the use of these mushrooms to benefit mental health.

>> **Medicine:** The first antibiotic used to fight human infections was called penicillin after the *Penicillium* mold that makes it (Chapter 14). Today, scientists are testing fungi like turkey tail (*Trametes versicolor*) for their ability to fight cancer and using fungi to produce cholesterol-lowering statin drugs.

Chapter **2**

Exploring the Role of Fungi in Nature

I f you look at a forest or a field, you probably notice the plants and maybe some animals. If you look closely, you may also see a few fungi, like shelf fungi growing on a tree or some mushrooms sprouting up through the grass. What you don't see is what is happening below the surface of that forest or field. Woven throughout the soil are the fine threads of growing fungi, making up as much as 90 percent of the living component of the soil by weight. These threads are busy absorbing water and digesting organic matter. Some of them even form partnerships with the plant roots around them. Fungi are in the environment all around us, but they often go unnoticed until they make a larger reproductive structure like a mushroom. This chapter presents what fungi are, as well as what they aren't, and takes a look at some of the ways they impact life on Earth.

Why Fungi Aren't Plants (and Never Were)

It's easy to understand why people might think fungi are plants. The visible fungi, such as mushrooms, form structures that don't seem to move or make noise, and they're often found growing among or even on plants. You can even walk up and pick a mushroom just like you'd pick a flower.

REMEMBER

Today, the term *mushroom* most often refers to the large, fleshy, reproductive structure of a fungus. This definition would include both the classic umbrella-shaped fungi and the shelf-like bracket fungi that you sometimes see growing on the side of wood. However, some scientists use an older, more restrictive definition of mushroom that only refers to the umbrella-shaped reproductive structure (also known as the fruiting bodies).

Appearances, as they say, can be deceptive. Even though fungi might look like plants, when scientists compared fungal DNA to that of other living things, they found that fungi are more closely related to animals (including you and me) than they are to plants. This wasn't a complete surprise to scientists, though, because when you look below the surface, you learn that the fungal lifestyle is very unlike that of plants.

Absorbing food from others

One of the biggest differences between fungi and plants is the way fungi get their food. Plants do photosynthesis, using energy from the sun to combine carbon dioxide and water to make sugars. Fungi get their energy and matter like we do by eating food.

REMEMBER

Plants are *autotrophs* (*auto*=self, *troph*=feed), which means they make their own food. Fungi are *heterotrophs* (*hetero*=other) because they must eat food that was originally made by other organisms.

Although fungi and animals are both heterotrophs, most fungi grow by microscopic threads called *hyphae,* like those shown in Figure 2-1. Hyphae grow and branch, forming a woven mass called a *mycelium* that makes up the body of the fungus. The fungal cells that form the hyphae have rigid cell walls. Because of their small size and cell walls, fungi can't ingest or engulf large food particles. (For more details on hyphae and fungal cell walls, check out Chapter 3.)

FIGURE 2-1: Mushrooms above the soil showing their masses of branching, thread-like hyphae growing through the soil.

REMEMBER

Fungi are *absorptive feeders*. They weave their hyphae through their food, releasing digestive enzymes outside of their cells. These *exoenzymes* help break large molecules down so that the fungi can absorb the smaller components into their cells.

In other words, if fungi were people, they'd be doing digestion all over the surface of their skin instead of inside their digestive systems. If you've ever seen a piece of fruit turned to watery mush by a mold, you've seen fungal digestion in action.

Creating a kingdom for fungi

People have been aware of and using fungi for thousands of years. People in Central America carved stones into mushroom shapes as early as 1000 BCE (Before Common Era), and edible mushrooms appear in art from the ruined city of Pompeii that dates to 79 CE. Books written on herbals during the Middle Ages contain descriptions and uses for both plants and fungi. But although people were using fungi for food, medicine, and spiritual rituals, they didn't really know what fungi were or where they came from. Some people saw them as plants, while others thought they had a supernatural origin. Several ancient cultures, for example, thought mushrooms appeared after thunderstorms.

The invention of the microscope led to more detailed studies of the structure of fungi. In 1665, an English doctor named Robert Hooke published a book of drawings that included the drawing of a mold as he saw it through a microscope (see Figure 2-2). His drawing shows hyphae and round structures called *sporangia* that contain *spores*, which are the reproductive cells of fungi.

Over the next few hundred years, scientists advanced their understanding of the genetics and reproduction of fungi, but most of them still put fungi in the same category as plants. It wasn't until 1969 that Robert Whittaker proposed that the fungal lifestyle was

FIGURE 2-2: Drawing of a "hairy mold" colony (later identified as the mold *Mucor*) published by Robert Hooke in 1665.

so different from that of plants that they needed their own category, which he called Kingdom Fungi. Whittaker placed fungi into their own kingdom because of their important role in breaking down dead organisms in the environment, their unique cellular structures, and the fact that they are heterotrophic, absorptive feeders. He also noted that most fungi are multicellular.

The study of fungi is called *mycology* (*myco*=fungus, *logy*=study of).

Mycologists and other scientists use several different terms when describing fungi:

>> *Fungus* (or plural fungi) refers to all organisms within Kingdom Fungi.

>> *Mold* is a descriptive term that refers to things that grow as multicellular filaments. *Mildew* refers to molds that grow on surfaces such as walls. Neither of these terms indicates a specific species or relationship.

>> *Yeast* is a descriptive term for fungi growing as single cells. Some fungi can switch between growing as a yeast to growing as a mold.

Since the 1960s, comparisons of DNA sequences between organisms have confirmed that fungi are distinct from plants and are actually more closely related to animals. Scientists often compare the genes for a molecule called ribosomal RNA (rRNA) to examine relationships between organisms. All cells have rRNA, so it's possible to use these comparisons to build a tree of all life on Earth, like the one shown in Figure 2-3, where groups of different types of organisms are shown as branches. Scientists use computers to draw the trees so that the physical distance along the branches represents the relationship between groups. The closer the two branches are, the closer the relationship.

Kingdom Fungi belongs to a large category of organisms called Domain Eukarya. This domain also contains the kingdoms Plantae (plants) and Animalia (animals), as well as many groups of microorganisms. All members of Domain Eukarya have similar cellular structures, which are discussed in Chapter 4.

Scientists are still working to understand the relationships between organisms within the fungal kingdom and to sort out the relationships between fungi and other closely related organisms.

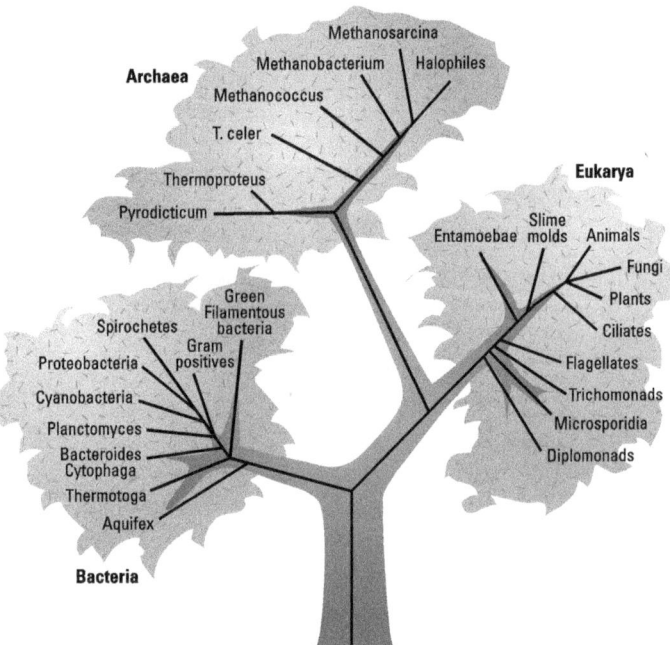

FIGURE 2-3: A phylogenetic tree of life based on comparison of rRNA genes.

This effort is complicated by the fact that we've only identified a very small fraction of the fungi on our planet.

REMEMBER

Scientists have identified about 150,000 different species of fungi, but they estimate that somewhere between 2.2 and 3.8 million species exist on Earth.

Fungi spend most of their lives growing microscopically, which is part of the reason that so few have been discovered and named. With so many species yet to be studied, it's likely that our understanding of the relationships between fungal groups will continue to change in the near future. In the meantime, mycologists will continue to study both fungi and fungal-like organisms.

Decomposing the Dead

Fungi may be the most underappreciated life form on Earth. They work quietly all around us, breaking down the bodies of the dead. Without them, we'd be surrounded by piles of every leaf that ever fell, every animal that ever died. (Imagine how quickly your home would fill up with trash if the garbage collectors stopped coming.)

REMEMBER

Most fungi are *decomposers,* organisms that break down dead material as a source of food. Scientists also refer to them as *saprobes* or *saprophytes* because they live on and break down decaying organisms.

WARNING

Some people think that the dead just break down or that worms and insects eat the dead. Dead bodies would not decompose without the action of decomposers like fungi and bacteria. Mummies or preserved bog people demonstrate that — if you block microbial decomposition (with chemicals or an acid environment) — bodies do not decompose. When an organism dies, bacteria and fungi begin the decomposition process. If the dead organism is an animal, the smell of decay attracts insects like blow flies and beetles that lay their eggs in the corpse. The eggs hatch, and the larvae feed on the decaying corpse. So, while insects are important contributors to the decay of some organisms, they are considered *scavengers* because they are animals that eat the dead.

Employing enzymes

Fungi are absolute rock stars when it comes to breaking things down. They produce a wide variety of digestive enzymes, enabling them to break down complex molecules into their component parts. These include enzymes similar to those made by your own digestive system, such as lipases to break down fats, proteases to break down proteins, and amylases to break down starch. Fungi take digestion even further with enzymes that break down the tough molecules found in plant cells, allowing them to digest materials that other organisms can't, like wood.

TIP

Enzyme names typically end in *-ase* and often refer to their function. For example, a*mylose* is a type of starch, so *amylase* is an enzyme that breaks down amylose.

REMEMBER

People use fungal enzymes for industrial processes, such as the production of paper, textiles, and biofuels. In fact, more than half of the enzymes used in industry today come from fungi. (For more on the use of fungi in biofuel production, head to Chapter 12.)

Plants produce a strong layer around their cells called a *cell wall* that can be difficult for other organisms to digest. In fact, you've probably heard of plant cell wall materials referred to as the *fiber* in your diet that passes through your digestive system without being broken down. When fungi find these fibers in animal feces or dead plant material, they represent an excellent source of energy and building materials:

>> All plant cells produce a *primary cell wall* that is flexible and surrounds growing cells. Primary walls contain the complex carbohydrates cellulose, hemicellulose, and pectin.

>> Some plant cells produce a *secondary cell wall* that is thicker and stronger than the primary cell wall. Secondary cell walls contain lignin in addition to cellulose and hemicellulose. Lignin is a complex branching molecule made up of alcohol subunits. It is what gives strength to plant cells, such as those found in wood.

- *Lignin* is a complex web of alcohol subunits.

- It's what gives wood its tough, rigid nature. Surrounds the cellulose and protects it from microbial decay.

Wood is one of the most amazing materials in nature. It's a complex mixture of several types of molecules, but the main components are cellulose, hemicellulose, and lignin. It's sturdy, rigid, and resistant to decay, which is why so many groups of people learned to use it for building material. For fungi, it presents an interesting problem because the useful carbohydrates are locked behind a wall of protective lignin.

Some fungi have developed strategies to deal with the lignin problem:

>> Fungi that cause *white rot* tackle the lignin head-on, breaking it down with enzymes called peroxidases. They almost totally break down wood, turning the wood white as they first digest lignin and then attack the cellulose. Wood decomposing by white rot turns white and often appears stringy or spongy.

>> Fungi that cause *brown rot* have developed a much more subtle approach. They use small ions (charged particles) to loosen the lignin and then sneak the carbohydrates out. They leave the lignin behind, causing the decomposing wood to appear dark brown and crumbly. If you've ever been hiking and reached out to touch a decaying tree stump and had pieces just break off in your hand, you were probably seeing the remains of brown rot.

Recycling carbon

The importance of fungi as decomposers isn't just about the clean-up of life's trash; it's about their importance as master recyclers. As fungi decompose the remains of the living, they release nutrients back into the environment for living things to use again. They break down the fats, proteins, carbohydrates, and DNA from the dead, using what they need for energy and building materials and releasing the things they don't need as waste. Fortunately for the rest of life on Earth, this means they release necessary nutrients like nitrogen and phosphorus back to the soil and carbon back to the air.

REMEMBER

Scientists track the movement of the elements essential to life, noting how they change forms as they pass through living things and the environment. They call these pathways *biogeochemical cycles* to represent the fact that the elements (chemicals) move through the living (biotic) and geological components of ecosystems.

The carbon cycle, shown in Figure 2-4, may be the most important biogeochemical cycle on Earth. Not counting water, carbon is the most abundant element in living things. It makes up the backbone of all of our big molecules: carbohydrates, proteins, lipids like fats, and nucleic acids like DNA. To build new molecules, cells, and organisms, living things need a source of carbon. Plants and other autotrophs can capture carbon dioxide out of the environment and turn it into the molecules that everyone else needs, but if we didn't have decomposers like fungi and bacteria, all of that carbon would eventually get locked up in the bodies of the dead. It's the decomposers that eat the dead, returning carbon to the air as carbon dioxide.

Photosynthesis and cellular respiration are two of the most important processes that affect the carbon cycle:

>> During *photosynthesis,* autotrophs use energy from the sun to combine carbon dioxide (CO_2) and water (H_2O) into carbohydrates like glucose ($C_6H_{12}O_6$). Oxygen gas (O_2) is produced as waste.

>> During *cellular respiration,* organisms use oxygen gas (O_2) to break down glucose ($C_6H_{12}O_6$), capturing usable energy and producing carbon dioxide (CO_2) and water (H_2O) as waste.

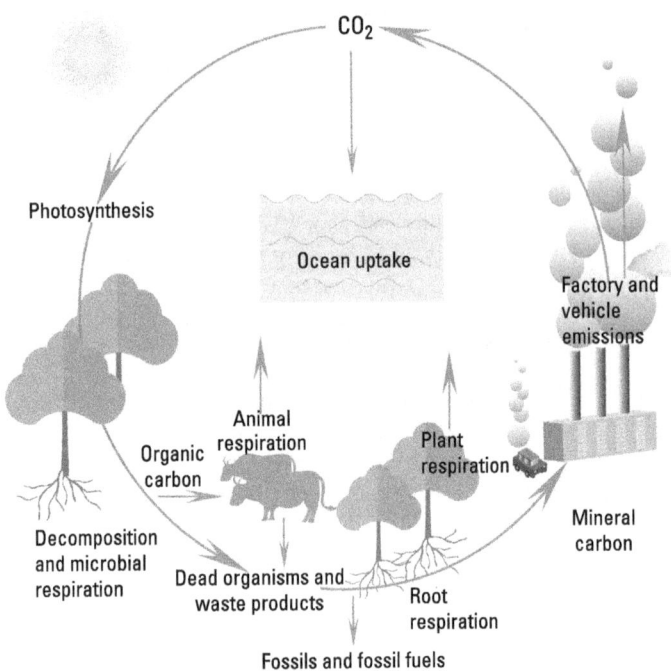

FIGURE 2-4: The carbon cycle.

danylyukk/Adobe Stock Photo

REMEMBER

The summary reaction for photosynthesis is $6\ CO_2 + 6\ H_2O +$ light energy $\rightarrow C_6H_{12}O_6 + 6\ H_2O$. The summary reaction for cellular respiration is $C_6H_{12}O_6 + 6\ H_2O \rightarrow 6\ CO_2 + 6\ H_2O +$ energy.

Although most organisms perform cellular respiration to break down food molecules, releasing some CO_2 back into the environment, the cellular respiration performed by fungi and bacteria during decomposition is essential to retrieving the carbon from the dead.

WARNING

Sometimes people think that when decomposers break down the dead, they release carbon back into the soil. While some mineral forms of carbon can be found in the soil, this is not where plants get their carbon, nor is it how decomposers release carbon back into the environment. Decomposers release carbon as carbon dioxide back into the atmosphere, and plants reclaim it from there.

The carbon cycle is also affected by physical processes such as forest fires, volcanic eruptions, and the burning of fossil fuels. People have been burning coal for thousands of years, but we dramatically increased our rate of consumption after we invented coal-burning engines during the Industrial Revolution. Coal and other fossil fuels formed from the remains of photosynthetic organisms that died but decayed very slowly due to the temperature, pressure, and oxygen concentrations at the time. Instead of being decomposed into carbon dioxide, the decaying organisms were transformed into fossil fuels. Now, as we rapidly burn these fuels, we're releasing that carbon back into the atmosphere as carbon dioxide at rates unprecedented in human history.

Carbon dioxide is a greenhouse gas, which means it can trap heat in the atmosphere, as shown in Figure 2-5. The large amounts of carbon dioxide we're putting into the atmosphere contribute to the warming of the Earth. The increased global temperature is having many significant impacts, including the melting of polar ice and rising ocean levels, droughts and disruptions to agriculture, higher risk for fires, and more severe weather systems.

The Greenhouse Effect

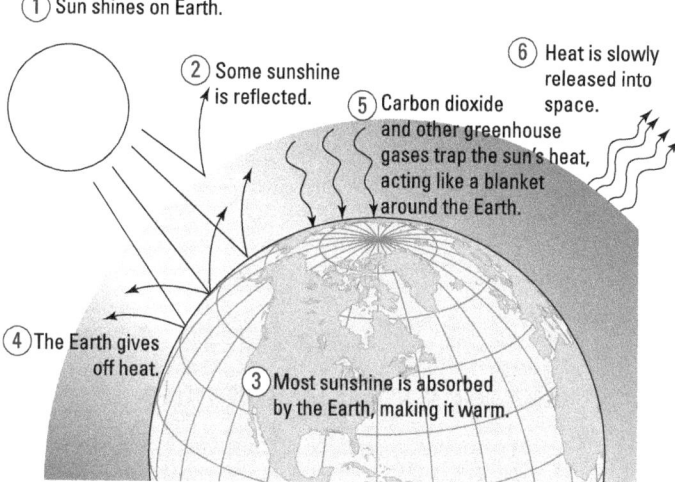

FIGURE 2-5: The greenhouse effect.

WHY DIDN'T FUNGI STOP COAL FORMATION?

Peak coal formation occurred during the later part of the Carboniferous Period, which extended from about 359 to 299 million years ago. Tree-like plants related to today's club mosses grew rapidly in tropical swampy forests. When they died, they fell into waterlogged soils, where they were transformed into peat. Marine sediments eventually washed over them and heat and pressure converted them to coal, trapping all of their carbon in the ground.

If you were to go outside today and find some dead plants, chances are they're currently being decomposed by fungi and bacteria. As they break down the plants, these decomposers release carbon into the atmosphere as carbon dioxide. So, why didn't this happen during the Carboniferous Period? How is it possible that so much dead plant matter was buried rather than completely decomposed?

Scientists have been studying this issue for a long time, and they've come up with many reasons. For one thing, plant growth during this moist, tropical period was very rapid. Also, decomposers grow more slowly in very wet soils due to the lower amounts of oxygen. This led to a situation where plants were growing and dying faster than the rate of decomposition. Some scientists have hypothesized that lower numbers of fungi able to degrade lignin during this time may also have contributed to the slower rates of decomposition. Studies of fossil fungi and fungal enzymes support the idea that some groups of fungi that can digest lignin didn't evolve until after this period of peak coal production. The fossil record of fungi is pretty limited, however, and more study needs to be done to determine if fungal evolution had a significant impact on coal formation.

Fighting global warming

Earth's average surface temperature has risen 2 degrees Fahrenheit (1 degree Celsius) since the beginning of the Industrial Revolution, with 2023 and 2024 both ranking as the hottest years on record, according to scientists from the U.S. National Oceanic and Atmospheric Administration (NOAA). Most scientists agree with the Intergovernmental Panel on Climate Change (IPCC) that human activity is the main reason for the change in temperature.

One of the biggest impacts is from the burning of fossil fuels, which releases greenhouse gases like carbon dioxide (CO_2) into the atmosphere. Greenhouse gases trap heat as it rises from the planet's surface, preventing it from escaping into outer space.

Deforestation also contributes to global warming. Plants remove carbon dioxide from the air and build it into their bodies, so forests represent carbon sinks.

A *carbon sink* is anything that absorbs more carbon dioxide than it releases, acting as a storage area for carbon.

What scientists are realizing is that it's not just the plants that help hold carbon in forests, it's the fungi too. Mycorrhizae are fungi that weave through the soil, forming an association with plant roots (see Chapter 3 for details). The fungi help plants access water and minerals, and the plants give some of the sugar they make to the fungi. That means that some of the carbon dioxide that plants remove from the air ends up building the mycelia of the mycorrhizae in the soil, creating another massive carbon sink (see Chapter 16 for an example of how big a soil fungus can get).

Parasitizing the Living

Unfortunately for some organisms, fungi don't always wait until you're dead to try to eat you. Fungi infect many species of plants and animals. Fungi called rusts and smuts, discussed in Chapter 11, cause significant agricultural losses. One fungus that causes Dutch elm disease has killed millions of elm trees, and another has pushed the American chestnut tree to the brink of extinction. Some fungi attack ants and spiders, turning them into zombies (more on this in Chapter 16) and some cause human diseases. In the next sections, you take a closer look at how fungi affect living hosts.

A *parasite* is an organism that lives in another organism for a significant period of time, getting what it needs to survive while harming the health of the host. A *pathogen* is a microorganism that causes disease in its host.

Attacking insects and plants

Even though animals and plants represent very different environments for fungi to live in, certain groups of fungi have produced species that use similar strategies to be successful in both types of organisms. Some parasitic species depend completely upon their host organism, while others can switch between a parasitic lifestyle or a free-living one. Some species have a very narrow range of hosts they can infect, while others can infect many different organisms.

Fungi that require a host are called *obligate parasites*. Those that can switch between living on decaying material and being parasitic are called *facultative parasites*.

The fungi that attack animals and plants usually begin their attack when a reproductive cell called a spore lands on and sticks to the host. When environmental conditions are favorable, the spore will *germinate* and begin to grow a hypha. Often, these hyphae develop special flattened tips called *appressoria* that use mechanical force to punch through the outer layer of the host, allowing access to the inside. The pathogens may also produce enzymes that break down the outermost layers of the host, making them easier to penetrate.

Fungi that attack insects are called *entomopathogenic fungi*.

Approximately 1,000 different species of fungi infect insects. (Some of these are so good at their jobs that people use them to control insects that attack crop plants.) Entomopathogenic fungi are different from other insect pathogens because they don't need to be eaten by the insect to cause disease. After they punch through the insect exoskeleton and reach the body cavity, they begin to grow, often switching from hyphal growth to growth as single-celled yeasts. Some insect pathogens produce toxins that kill their hosts quickly and then continue to digest their remains. Others keep their hosts alive and may even control their behaviors in ways that optimize fungal reproduction. In the end, the fate of the insect is the same: the insect is digested, and the fungus reproduces, releasing new spores into the environment.

Fungi are the largest group of plant pathogens. Scientists estimate that between 20 and 40 percent of the potential global food harvest is destroyed by fungi. Some of the greatest famines in history, including the Irish potato famine, were caused by fungi or fungal-like organisms. In addition to causing food shortages by attacking crops directly, fungi are also responsible for the loss of stored food due to rot and deterioration.

Threatening the survival of amphibians and bats

The impact of parasites on host populations can be devastating when a host species has little resistance. This often occurs when a parasite is first introduced to a new host species. Unfortunately, global circulation of people and goods increases the chances that this may occur. When people travel, they track microbes with them. The global trade in animals for pets not only moves pets, it also moves their pathogens. In many cases, the movement of microbes has given parasites deadly access to new host populations.

Driving amphibians to extinction

Scientists estimate that almost 41 percent of the world's amphibian populations are currently threatened with extinction, and as many as 222 species may have gone extinct in the last 150 years. The primary cause of extinction is habitat loss due to human activities such as agriculture, logging, and development. Added to these pressures is death caused by the invasive fungal pathogen *Batrachochytrium dendrobatidis*.

B. dendrobatidis belongs to a group of fungi called chytrids (discussed in Chapter 8) and causes a disease called chytridiomycosis. This pathogen has been found all over the world but is having the biggest impact on amphibian populations in South and Central America, Australia, and North America. The fungus invades the skin of frogs and other amphibians, possibly affecting their ability to exchange oxygen through their skin. Infected frogs may start shedding their skin, which turns a dull greyish color as the fungus spreads through it. When the fungus reproduces, it releases a swimming stage that can move through water to find new hosts.

TIP

People can help prevent the spread of this disease by cleaning boots, clothing, and equipment before moving between bodies of water and by never moving amphibians from one area to another.

SURVIVOR

Imagine you are high in the Sierra Nevada mountains in Yosemite National Park. Scattered around the beautiful vista are small, jewel-like lakes. And when you approach those lakes, little yellow-legged frogs leap from both sides of the shore, creating a visual shower of frogs as they return to the lake. If you hiked into the high Sierras and saw this sight today, you would be seeing something that almost disappeared from Earth.

The Sierra Nevada yellow-legged frog once filled the fish-free lakes of the high Sierras, providing food for bears, birds, coyotes, and snakes. But when Europeans came through the mountains on their way to find gold in the 1800s, they wanted to be able to fish for food. They began introducing non-native fish that were familiar to them. The fish ate the tadpoles, and the frogs were lost from all but the most remote lakes. Then, after World War II, people began using airplanes to drop fish into even the remote lakes, and it looked like the frogs' days were numbered. The frogs got a second chance when Yosemite National Park stopped stocking lakes with fish in the 1990s.

In 1992, scientist Roland Knapp got permission to try and help the frog populations recover by removing non-native fish populations from some lakes. It worked, and the frog populations began to bounce back. Agencies that manage the wildlife in that area followed his example, and it looked like the yellow-legged frog was on its way back to its former glory.

But then, in the early 2000s, the fungus *B. dendrobatidis* that was killing frogs around the world found its way to Yosemite. The yellow-legged frogs started dying and the population in lake after lake crashed. Scientists listed the frog as an endangered species in 2014.

Just when scientists had almost given up hope, they noticed that some frogs were surviving the fungal disease and that the populations of some lakes were starting to recover. It seemed that some of the yellow-legged frogs were resistant to infection by the fungus. The scientists started capturing some of these frogs and moving them to new lakes, in the hopes that they would be able to survive and help the frog populations recover. Their efforts were successful, and today you can indeed hike up into the high Sierras to see a sight that would have disappeared if it weren't for the help of some dedicated scientists and the resilience of one tiny frog species.

Killing bats with white-nose syndrome

Since 2006, bats in North America have been under siege by white-nose syndrome (WNS), caused by the fungus *Pseudogymnoascus destructans*. The disease gets its name because the fungus becomes visible as white fuzz on the nose and wings of bats during the later stages of the disease. Scientists don't know how the fungus came to North America, but they have found it living among European bats in whom it doesn't cause severe disease. The global travel of people, animals, and merchandise makes it very possible that reproductive cells called spores were carried from Europe to North America, allowing them to infect new populations of bats.

The fungus spreads when bats come into contact with spores, either by touching other bats or by touching cave walls with spores on them. The fungus grows best in cooler temperatures, making the inactive bodies of hibernating bats a prime target. The fungus damages the bats' wings and makes it harder for them to regulate their water balance. Hibernating bats that are infected with the fungus use twice as much energy during the hibernation season and may starve to death as a result.

Millions of North American bats have died since the fungus appeared in 2006. Populations of the little brown bat, the Northern long-eared bat, and the tri-colored bat have declined by 90 percent. Bats like these are important to their environment because they control lots of insect populations, including mosquitoes and some agricultural pests. Wildlife managers are trying to save the bats by controlling the movement of people and materials between areas affected by WNS and those not affected. For example, if you explore caves in an area where WNS is known to affect bats, you should disinfect your boots and clothing before going into caves where WNS isn't present. Scientists are testing to see whether vaccination can help protect bats and monitoring what factors seem to be associated with greater survival.

Causing diseases in people

Approximately 200 species of fungi are known to infect humans, causing a variety of conditions that range from superficial infections of your skin or nails to more serious infections below your skin to life-threatening deep infections of your organs. One of the challenges in fighting fungal diseases is that their cell structure is very similar to ours, so many chemicals that might stop a fungus from growing would also harm a person.

Discovering what grows between your toes

The fungi that infect human skin are called *dermatophytes,* which literally means skin (*derma*) plants (*phytes*). Many of these fungi make enzymes that help them break down keratin, the tough protein that reinforces our skin, hair, and nails. They grow best in moist areas of the body, such as between the toes, around the groin, or in between folds of skin, where they cause a variety of conditions, including athlete's foot, jock itch, and ringworm.

REMEMBER

Mycoses (*mycosis,* singular) are fungal diseases. Infections of the skin are called *cutaneous mycoses* or *tinea.* Tinea is usually followed by the location of the infection, for example, tinea pedis means a skin infection of the foot.

Athlete's foot is the most common fungal skin infection in people, affecting about 70 percent of people at some point in their lives. The signs and symptoms include itchy, dry, peeling, or cracked skin on the feet. Ways to minimize your risk for athlete's foot include wearing dry cotton socks and protecting your feet in public areas.

The fungi that commonly cause athlete's foot, *Trichophyton* and *Epidermophyton,* also cause other forms of skin and nail infections. Athletes who participate in contact sports are at higher risk for ringworm, which can appear almost anywhere on the body. It gets its name from its appearance on the skin as a flat circular patch surrounded by a rough, red border. Ringworm is caused by the fungus *Microsporum.* The dermatophytes that cause these diseases belong to the group of fungi called Ascomycota (discussed in Chapter 10).

REMEMBER

Ringworm is named for the red ring that can appear on the skin. It's caused by a fungus, not a worm.

WARNING

Although fungal skin infections aren't serious for most people, they can be a more serious problem in people with diabetes or people who are immunocompromised.

Considering serious infections

When fungi invade the interior of the body, they can cause serious, even fatal, diseases. The most common way that fungi invade the body is when people inhale fungal spores. Healthy people don't usually get infected as a result, but people with weakened immune systems can develop serious lung conditions like pneumonia. In people with late-stage HIV disease (AIDS), fungal lung infections may be fatal.

Here are some of the conditions that can result from the inhalation of fungal spores:

>> **Histoplasmosis** is a lung infection that results from breathing in spores of the soil fungus *Histoplasma capsulatum*. Healthy people don't usually develop the disease, but people with weakened immune systems can develop chronic lung conditions, life-threatening pneumonia, or meningitis, which is a dangerous swelling of the lining around the brain.

>> **Valley fever** (coccidiomycosis) results from breathing in the spores of *Coccidiodes*, a fungus that lives in the soil of the Pacific Northwest and southwestern United States, as well as Mexico and parts of Central and South America. Some people who breathe in the spores don't get sick, but other people will develop a lung infection with cough and fever. Rarely, the fungus will spread to other parts of the body.

>> **Cryptococcosis** is a serious infection resulting from inhalation of the spores of *Cryptococcus*. In people with weakened immune systems, it can lead to a lung infection or cryptococcal meningitis. Symptoms of a lung infection include fever, cough, and chest pain, while meningitis leads to headache, neck pain, light sensitivity, and confusion. Cryptococcosis can be life-threatening and needs to be treated with antifungal medication.

>> **Aspergillosis** is an infection of the lungs that results from inhalation of the spores of *Aspergillus* mold. Healthy people don't usually develop this infection, but people with weakened immune systems or other lung diseases may experience coughing and shortness of breath. The species *A. fumigatus* has developed antimicrobial resistance, making it harder to treat.

>> **Talaromycosis** typically occurs in people with weakened immune systems who live in or visit Southeast Asia or Northeast India. It's caused by the fungus *Talaromyces marneffei*. Initially, the infection results in small red bumps on the skin, but these bumps can spread to other areas of the body, especially in people with advanced HIV disease (AIDS).

>> **Pneumocystis pneumonia** is a rare, serious lung infection caused by *Pneumocystis jirovecii* that usually develops in people with weakened immune systems. About 30 to 40 percent of people who develop pneumocystis pneumonia have advanced HIV disease, and it is a leading cause of AIDS-related mortality.

Finding a magic bullet to fight fungal infections

The biggest challenge in fighting fungal infections is that their cells are so similar to ours. Chemicals that stop fungal growth or kill fungal cells are likely to damage our cells, too, and have dangerous side effects. To treat fungal diseases, scientists must find drugs that target the small differences between the cells of fungi and those of the human body.

TECHNICAL STUFF

Many antifungal drugs target a molecule called *ergosterol* that fungi need to build one of the boundaries of their cells, called a cell membrane. Because our cells don't make ergosterol, we can use medicines that target this molecule without hurting our own cells. Some older antifungal medications contain the compound amphotericin B, which binds directly to ergosterol and disrupts fungal membranes. Several newer and safer drugs target various fungal proteins that help manufacture ergosterol. Treatment for fungal skin infections usually involves antifungal creams or sprays. Treatment for invasive fungal infections requires taking these medications internally.

Making Connections in Communities

Fungi are major components of most ecosystems on Earth. They interact with other organisms in various ways, from helping recycle nutrients through decomposition to forming beneficial partnerships with plant roots and algae (see Chapter 3) to parasitizing insects, plants, and even other fungi. Through these connections, they make major impacts on the ecosystems in which they live.

Microbes in the environment often grow in *biofilms*. Biofilms form when microbes attach to a living or non-living surface and produce and grow within a slimy matrix. Entire communities of organisms can grow within the same biofilm, exchanging materials and communicating with each other.

Fungi grow in a variety of biofilms, from those on the surfaces of living plant roots to those on dead plant material. Pathogenic fungi form biofilms during the development of disease in both plants and humans, so biofilm formation represents another potential target for antifungal drugs. Fungal biofilms associated with plant roots may include beneficial soil bacteria and have positive impacts on plant growth.

Chapter **3**

Benefitting from Fungal Relationships

L iving things depend on each other to survive. Some relationships are obvious, such as those between predators and their prey. Other interactions are more subtle, even invisible. If you looked in a mirror, you would see a person. What you wouldn't see are the microbes living in your digestive system that provide you with vitamins. Similarly, if you look outside, you can see plants. What you may not see are the fungal partners that help those plants grow. In this chapter, you take a look below the surface and see how fungi partner with other organisms to help them grow.

Exploring the Wood Wide Web

When you take a walk through a forest, you can see the trees all around you, and you may spot some mushrooms on the forest floor growing among the trees. On the surface, the trees and mushrooms may seem like neighbors that grow near each other but remain separate individuals. If you took the time to look beneath the forest floor, however, you would realize that the connections between the trees and mushrooms are so much more. The mushrooms you see are just the visible tip of a huge fungal network that spreads throughout the soil and connects with plant roots (see Figure 3-1).

The fungi pass molecules around this network, giving minerals to plants and receiving sugars in return. Some research suggests that mycelial fungi may even pass materials from one tree to another, helping the trees communicate with each other. Scientists who study this vast network sometimes compare it to the Internet, where the trees are the nodes and the fungi are the connections between them.

This association between trees and fungi isn't casual or temporary. It's a long-term, stable partnership that persists throughout the lives of both organisms, sometimes for thousands of years (see Chapter 16 for more on the largest and possibly oldest fungus, *Armillaria ostoyae*). Scientists call a long-term partnership like this one a symbiosis.

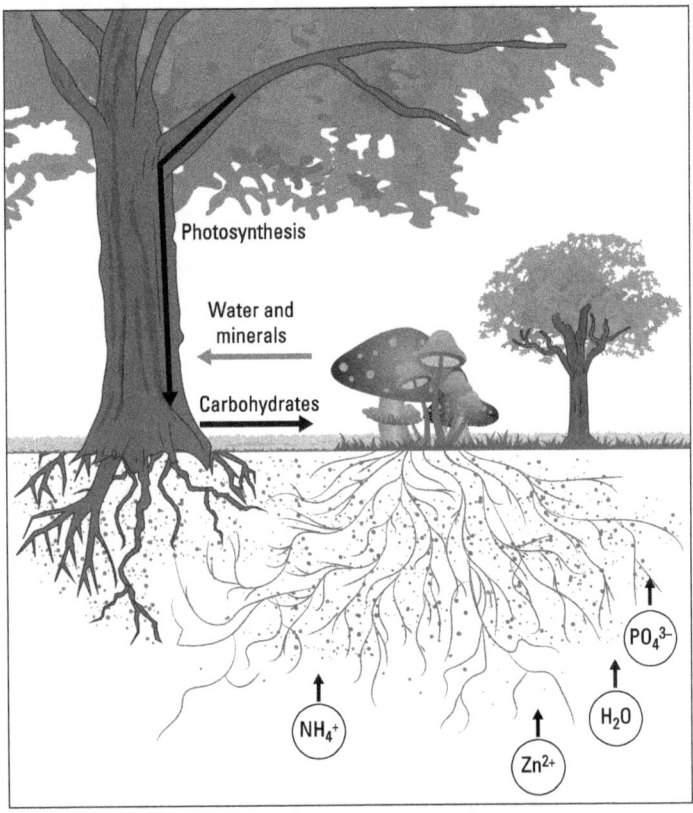

FIGURE 3-1: Symbiotic associations between plants and fungi.

REMEMBER

Symbiosis occurs when two organisms live together for a significant part of their life cycles.

WARNING

People commonly use the word symbiosis to indicate a positive relationship between two things. In scientific terms, however, symbiosis can be beneficial or harmful for the partners. When a symbiosis is beneficial to both partners, the symbiosis is a *mutualism*. When it benefits one but is harmful to the other, it's a *parasitism*.

Extending the reach of plant roots

The fungi that form intimate associations with plant roots are called mycorrhizae, which literally means "fungus root" (*myco*=fungus, *rhiza*=root).

REMEMBER

Mycorrhizae (singular=*mycorrhiza*) are symbiotic associations between fungi and plant roots.

Scientists estimate that 80 to 90 percent of all plants have a relationship with mycorrhizae. Only certain species of fungi form mycorrhizae, and each one forms partnerships with specific plants or groups of plants. Most scientists consider the symbiosis to be a mutualism that benefits both partners. Mycorrhizal fungi provide plants with many benefits:

>> **Increased surface area of plant-soil interface:** The fine tendrils of hyphae spread out into the soil. Every place the soil touches the hyphae is a surface through which water and minerals can be absorbed. Scientists estimate that mycorrhizae expand the reach of plants by a factor of 10 to 100 times.

>> **Increased solubility of minerals in soil:** Minerals in soil often cling to clay particles by weak electrical charges. Fungi can release charged particles like hydrogen ions (H^+), weakening the attraction of minerals to clay particles and making it possible for the minerals to be absorbed.

>> **Improved uptake of nutrients:** Fungi absorb minerals like nitrogen (N) and phosphorus (P) that plants need to grow, then pass those minerals through the hyphae to the plant roots. Mycorrhizae provide up to 80 percent of the nitrogen and phosphorus a plant needs.

>> **Protection against pathogens:** The presence of mycorrhizal fungi on plant roots can prevent fungal pathogens from successfully attacking the plant. Plants that have associations with some types of mycorrhizae are also better at resisting attack from small worms called nematodes in the soil.

>> **Exchange of carbon between plants:** Mycorrhizae can move carbon in the form of carbohydrates from one plant to another, sometimes between plants of the same species, but also between different species. Scientists speculate that older established trees, nicknamed "mother trees," may use the mycelial network to send carbohydrates to younger trees, helping them to survive and get established.

The main benefit mycorrhizal fungi receive from their plant partners is food. Plants do photosynthesis and share some of the sugars they make with their mycorrhizae. For example, scientists estimate that trees give 10 percent of their total photosynthetic product to their fungal partners.

Scientists who study mycorrhizae have discovered some very interesting behaviors that suggest the symbiosis may be more like a business relationship than a true romance. Some mycorrhizae hoard precious minerals, driving up demand so they can exchange the minerals for a higher carbohydrate payment. They also sometimes skip their nearest plant contacts in order to trade with plants farther away that are offering a better price. And, once the fungi receive the sugars, they convert them into different forms of carbohydrate that plants can't absorb, ensuring that the plants can't ask for a refund.

Scientists categorize mycorrhizae into two major types depending on whether the fungi actually penetrate into cells of the plant root (see Figure 3-2):

>> *Ectomycorrhizae* grow over the surface of plant roots, forming a sheath of hyphae called a *mantle* (*ecto*=outer). The mantle is visible as a coating of hyphae on the outside of the roots. Hyphae extend outward into the soil and inward into the root, but they pass in between the plant cells.

>> *Endomycorrhizae* penetrate into the interior of root cells (*endo*=inside). Hyphae aren't visible on the exterior of the root.

We speak for the trees: Ectomycorrhizae

Ectomycorrhizae dominate in coniferous forests and are also important in broad-leaved forests in temperate and Mediterranean climates. They associate mainly with woody plants, especially conifer trees like pine, fir, hemlock, and spruce, as well as woody flowering plants like birch, willow, beech, eucalyptus, and oak trees. One ectomycorrhizal fungus may be able to form associations with multiple plant species, creating the potential for interspecies connections. Most of the fungi that form ectomycorrhizae belong to the Basidiomycota (Chapter 11), but a few belong to the Ascomycota (Chapter 10). A few of the more familiar ectomycorrhizal basidiomycetes are morels, chanterelles, boletes, and brittle gills (*Russula*), but there are so many more. The most famous ectomycorrhizal ascomycetes are the truffles that grow in association with trees like oaks.

Ectomycorrhizae have three defining structures:

>> A *mantle* of hyphae that covers the root

>> A web of hyphae called the *Hartig net* that grows into the root by passing in between the cells of the epidermis and cortex of the root (see Figure 3-2)

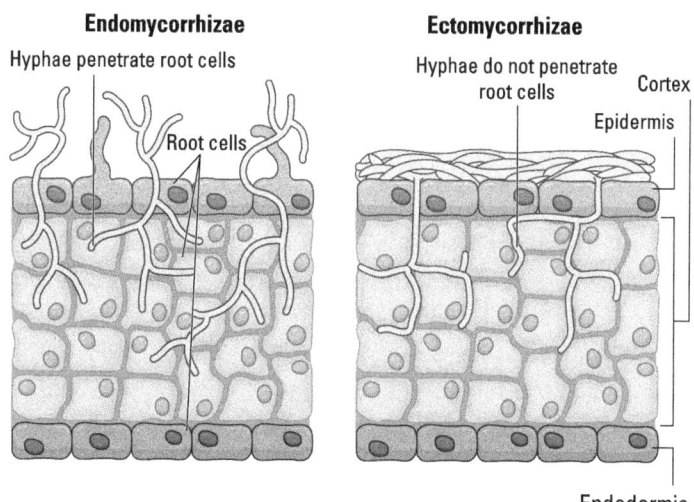

FIGURE 3-2: A comparison of ectomycorrhizae and endomycorrhizae.

>> *Extraradical hyphae* that grow outward from the root, forming a mycelial network throughout the soil and growing upwards to form reproductive structures like mushrooms at the soil surface

Establishing the partnership

Ectomycorrhizae can form on new roots growing from established plants or from newly germinating seeds. When new roots form from established plants, the new root is usually colonized by hyphae from the mycorrhizae on established roots. When a seed germinates to produce a new root, ectomycorrhizae can form either from contact with spores or mycelium in the soil.

The plant roots and the fungus interact with each other to establish a partnership. Plant roots produce compounds that may encourage spore germination and attract compatible mycorrhizal partners to grow toward them. The fungi produce plant growth regulators that change the way the roots grow, resulting in shorter roots that don't produce root hairs.

Forming nets around plant roots

Once the fungal hyphae recognize and attach to epidermal cells near the actively growing tip of a new root, the following steps occur:

1. The mycelium grows over the surface of the root, forming the mantle.

2. Hyphae grow outward from the mantle into the soil.

3. Hyphae penetrate the root by growing between and around root cells to form the *Hartig net*. The exchange of materials between the fungus and the plant occurs at the surface of the Hartig net.

   a. In angiosperms, the Hartig net forms between the epidermal cells of the root.

   b. In gymnosperms like pine trees, the hyphae penetrate more deeply into the root, forming the Hartig net in the cortex.

Getting preyed upon by mycotrophic plants

Some plants have developed a very un-plantlike way of life that takes advantage of ectomycorrhizae. These plants have given up their green and no longer do photosynthesis. Instead of making their own food, they take food from the mycorrhizal network. Ultimately, the source of this food is other plants hooked up to the same network.

REMEMBER

Mycotrophic plants get their name because they take carbohydrates from fungi instead of making their own (*myco*=fungus, *trophic*=feeding).

Many plant families have mycotrophic members. If you are ever out hiking and you spot a waxy-looking plant with pale white stalks and nodding white bell-shaped flowers, you may be looking at *Monotropa*. This plant's nickname is ghost pipe because of its pale appearance and resemblance to a small pipe. It's a member of the heath family, which also includes heather, blueberries, and cranberries.

Relying on arbuscular mycorrhizae for better plant growth

Arbuscular mycorrhizae are the most common type of mycorrhizae. They form between a group of fungi called the Glomeromycota (see Chapter 9) and a wide variety of plants, including flowering plants, some gymnosperms like *Sequoia,* ferns, mosses, and liverworts. The fungal partners can't grow independently, so scientists refer to them as *obligate symbionts.* Arbuscular mycorrhizae are defined by two characteristics:

>> They are endomycorrhizae that form structures within the cells of plant roots (see Figure 3-2).

>> Most species form structures called *arbuscles* inside plant cells. Arbuscles are branched clumps of hyphae that look like little trees (see Figure 3-2). Nutrient exchange between the plant and the fungus occurs at the arbuscle.

Many arbuscular mycorrhizae also form *vesicles,* which are sac-like storage structures. (In fact, the old name for arbuscular mycorrhizae was vesicular-arbuscular mycorrhizae.)

Arbuscular mycorrhizae are found in most ecosystems and are especially important in grasslands and tropical forests. They form with most crop plants, including wheat, rice, and corn, making them incredibly important to the human food supply. The fungi help plants access water and necessary nutrients like nitrogen, phosphorus, zinc, and copper. Plants with arbuscular mycorrhizae grow bigger and can resist plant pathogens.

Arbuscular mycorrhizae are also important for the general health of the soil. They produce sticky molecules that hold soil together and help keep nutrients in the upper layers where they are available to plants. Scientists who study sustainable agricultural practices say farmers could reduce their dependence upon chemical fertilizers if they adopted practices that encouraged the diversity and stability of arbuscular mycorrhizae in the soil.

Invading plant roots

The development of arbuscular mycorrhizae begins with an exchange of signaling molecules between the plant and the fungus, causing the fungal hyphae to grow toward the root. Once contact is made:

1. The fungal hypha forms a specialized attachment structure, called an *appressorium*, between epidermal cells at the surface of the root.

2. Hyphae grow from the appressorium and penetrate the cell walls of root cells, pushing the plasma membrane inward and growing between it and the cell wall.

   a. In some plants, the hyphae move between cortical cells until they reach an inner cell, then penetrate the inner cell wall and form an arbuscle.

   b. In other plants, the hyphae penetrate the walls of cortical cells closer to the outside of the root, then grow through cortical cells, forming arbuscles and twisting hyphae along the way.

Helping heather grow: Ericoid endomycorrhizae

Ericoid endomycorrhizae are mycorrhizae formed with plants in the Ericaceae family. This family includes heath (*Erica*), rhododendron (*Rhododendron*), blueberry and cranberry (*Vaccinium*), and wintergreen (*Gaultheria*). These plants typically grow in acidic soils that have low nutrient availability and require mycorrhizae to survive. The fungi that form these partnerships are mostly ascomycetes (Chapter 10) and some basidiomycetes (Chapter 11) that have the ability to survive in acidic soils. They produce enzymes that break down dead organic matter, playing an important role in nutrient cycling within these ecosystems.

The fungi make contact with the root hairs of the plants and then cover them with a loose layer of hyphae. This layer is not as dense as the mantle of the ectomycorrhizae. From this layer, hyphae penetrate through the cell wall of cortical cells and form coiled structures called *pelotons* (similar to the ones shown in Figure 3-3 in orchidaceous endomycorrhizae). These structures are the site of nutrient exchange with the plant.

Getting orchids started: Orchidaceous endomycorrhizae

At some point in their lives, all orchids require support from a fungal partner. Some orchids only need help in between germination and the growth of their leaves. Orchid seeds contain very little stored food, so the mycorrhizal fungus provides the plant with food until the leaves grow enough to support photosynthesis. Most orchids gradually become more self-sufficient and shift to a more typical mycorrhizal relationship where the fungal partner provides water and minerals in exchange for carbohydrates. However, some orchids never grow green structures and remain dependent on their mycorrhizae for their entire lives.

Orchid mycorrhizae are basidiomycetes (Chapter 11). Once the orchid germinates, the fungi attach to the root hairs. The hyphae penetrate the cell walls of the orchid cells and form coiled structures called *pelotons* (see Figure 3-3) between the cell wall and plasma membrane of the orchid cells. The pelotons facilitate the exchange of nutrients with the plant for a few days, then they collapse and get digested by the orchid. Fungal hyphae can re-enter the cell and form new pelotons.

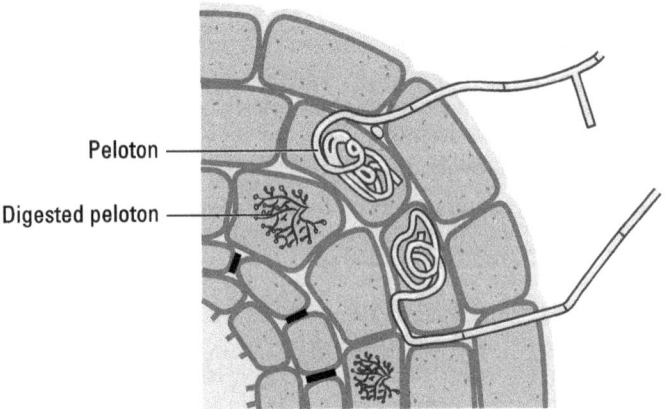

Peloton

Digested peloton

FIGURE 3-3: Orchidaceous endomycorrhizae.

Living on the Edge: Lichens

Lichens, like those shown in Figure 3-4, grow in places where other forms of life can't survive. If you've ever noticed spots of color on high rocks next to the ocean, for example, splashes of bright yellow, light green, or black that almost look like paint, you were probably noticing lichens. If you travel high into the mountains or into cold polar regions to the point that you leave behind the trees and then the smaller plants, chances are you'll still find lichens growing on the rocks and other surfaces. Lichens even grow on solidified lava flows near volcanoes. Lichens can grow in areas where other things can't because they can tolerate extremes of temperature and water loss. The secret to their superpowers? Teamwork.

A lichen is a stable partnership between two types of organisms: a fungal partner, called the *mycobiont,* and a photosynthetic partner, called the *photobiont* (also sometimes called the *phycobiont*).

Lichens form partnerships between certain fungi and algae. The algae may be eukaryotic species, such as green algae, or prokaryotic blue-green bacteria (also called cyanobacteria). The single-celled green alga *Trebouxia* is the photobiont in 75 percent of lichens. Most of the fungi that form lichens are members of the group Ascomycota (Chapter 10), but a few belong to the Basidiomycota (Chapter 11). Lichens can also form from partnerships between a fungus and two photobionts.

The fungal hyphae form the structure of the lichen thallus. Some lichens develop very organized bodies with distinct layers, as shown in Figure 3-5. Scientists call these lichens *heteromerous* lichens (*hetero*=different, *-mer*=part, so these lichens have different parts). In heteromerous lichens, the fungal hyphae weave a tight protective layer called a *cortex* at the top and bottom surfaces of the body. The photobiont lives in the *algal layer* just under the upper cortex, where it is protected but can get plenty of light for photosynthesis. Other lichens don't form organized layers. In these *homoiomerous* lichens (*homo*=same), the thallus is uniform, and the photobiont is scattered throughout the lichen.

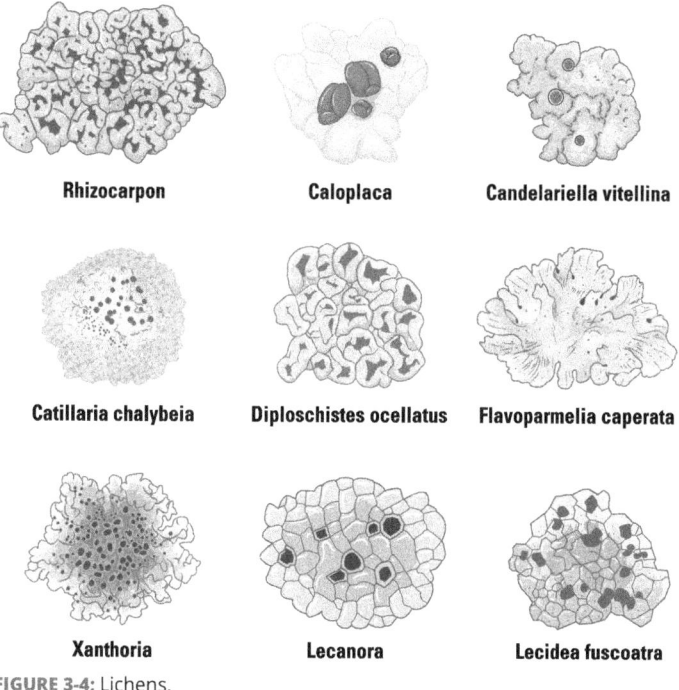

Rhizocarpon	Caloplaca	Candelariella vitellina
Catillaria chalybeia	Diploschistes ocellatus	Flavoparmelia caperata
Xanthoria	Lecanora	Lecidea fuscoatra

FIGURE 3-4: Lichens.

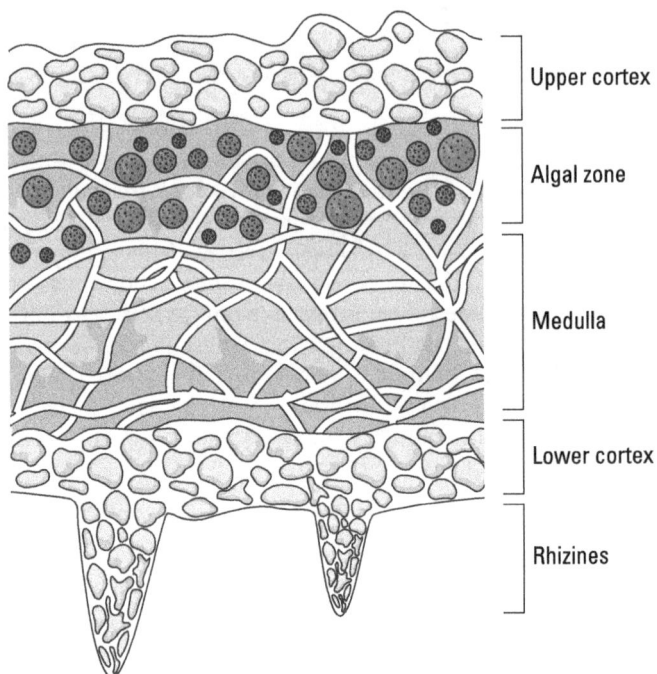

Upper cortex

Algal zone

Medulla

Lower cortex

Rhizines

FIGURE 3-5: The organization of a heteromerous lichen thallus.

Contributing to the partnership

The symbiosis between the fungus and its photosynthetic partner seems to be a mutualism, with both partners benefiting from the association.

REMEMBER

The photobiont does photosynthesis and shares sugars with the mycobiont. The fungus builds a house for the photobiont, protecting it from water loss and very intense light.

One way to measure whether a symbiosis is beneficial to the partners is to separate them and grow them independently from each other. If organisms grow better together than they do alone, then the symbiosis is a mutualism. The algal partners from lichens are fairly easy to grow in isolation, but it's proved more challenging to isolate and grow the mycobionts. Scientists who have successfully grown the mycobionts found that they grow more slowly than they do in the lichen and don't reproduce sexually. These results support the idea that the lichen partnership is beneficial for the fungus.

Some question whether the lichen is also beneficial for the photobiont. After all, they can grow just fine on their own, and they make their own food through photosynthesis. It's possible that the fungi are essentially holding the algae captive in a relationship that's more like a parasitism than a mutualism, and that the algae are throwing carbohydrates at the fungus in a desperate attempt to avoid being eaten. The strongest argument that the algae actually do benefit comes from the fact that lichens grow on land in all kinds of environments that the photobionts could never grow in on their own. Also, the photobionts can do photosynthesis within the lichen in conditions of lower water availability than they can on their own.

Lichens don't have any way to regulate their water content. When their environment dries out, they dry out. When a lichen dries out, it slows its metabolic rate and becomes dormant. In the dormant state, they can survive for months or years. When the environment has water, they hydrate and resume metabolic activity, including photosynthesis. It takes a lot to kill a lichen, but if one becomes dry for too long, or becomes very hot and dry, it may die.

Lichens are *poikilohydric*, which means they gain and lose water passively from their environment.

Lichens are well adapted to cycles of wetting and drying:

>> **They can absorb water from their environment even when very little is available.** Lichens absorb water directly through their outer layer or cortex. In many environments, lichens hydrate from the morning dew. They can also absorb water from fog. The ability of lichens to absorb water from low-water environments enables them to colonize extremely dry environments where plants can't grow.

>> **They produce protective molecules like polyols.** These molecules protect their cell membranes and proteins from the stress of drying out.

>> **They become dormant when dry, allowing them to conserve energy.** When they rehydrate, the first thing they do is start breaking down carbohydrates via cellular respiration. Once light is available, they begin photosynthesis.

Saving energy when they are dried out makes it possible for them to make enough food with shorter periods of photosynthesis.

>> **They can do photosynthesis even when their water content is low.** Even if they only have 25 percent of their maximum hydration, they can still make food.

>> **They can survive extremes of heat, cold, and solar radiation when they are dry.** This gives lichens a significant advantage in stressful environments like deserts and high mountains.

Lichens can survive harsh conditions in part due to their unique chemical compounds called *secondary products* or *secondary metabolites*. (*Primary metabolites* are the chemicals that organisms make for their basic structure and function, like carbohydrates from photosynthesis or cell wall materials.) Scientists have identified over 1,000 secondary products from lichens and are currently researching their properties and potential uses:

>> **Properties:** Secondary products protect lichens from UV radiation, helping them live in areas with intense light. They defend lichens from grazing animals, and they have antibacterial properties, which may help protect them from parasites. The amounts and types of secondary products produced vary in response to the type of substrate that the lichen grows on, so the chemicals may also play a role in substrate colonization.

>> **Potential uses:** Scientists determined that lichen products could kill some species of antibiotic-resistant bacteria, including MRSA (methicillin-resistant *Staphylococcus aureus*). Lichen products have also been successful at killing cancer cells, although none have yet been tested in patients. The potential power of these molecules is very exciting, but many research challenges need to be solved. Lichens grow very slowly, and they make these chemicals in small amounts. To harvest enough of a chemical for use as a medicine, scientists need to develop better systems for growing lichens in the lab.

Discovering the diversity of lichen shapes

Lichens vary widely in their appearance, from growths so flat they look like rust or paint on rocks, to leafy, three-dimensional structures easily mistaken for plants (see Figure 3-4). Although scientists who study lichens have many different terms for the appearance of different lichens, these can be combined into three main growth forms:

>> **Crustose** lichens grow as flat crusts on all kinds of surfaces, including rocks, trees, and buildings. These are the lichens that look almost like splashes of paint or odd minerals on otherwise bare rocks. Some examples of crustose lichens are *Rhizocarpon* and *Catillaria chalybeia* in Figure 3-4.

>> **Foliose** lichens have leaf-like lobes that aren't firmly attached to the substrate. As a result, they're often mistaken for plants and even given common names that include the word moss (true mosses are plants). For example, the oak moss in Figure 3-6 is not a moss but a lichen. *Flavoparmelia caperata* and *Xanthoria*, shown in Figure 3-4, also show foliose growth.

>> **Fruticose** lichens grow in all three dimensions, producing complex shapes like little shrubs, long strings, upright pillars, and little cups, like those of the pixie cup lichen in Figure 3-7. Many fruticose lichens are mistaken for plants and have common names like reindeer moss or Icelandic moss.

TIP

It can be hard to tell the difference between mosses and lichens, especially because they're often growing together on rocks, stumps, and trees. True mosses are plants that produce small leaves off a main stem. Lichens may look leafy, but you won't see distinct leaves and stems if you look closely.

Exploring the role of lichens in ecosystems

Lichens grow all over the world and in all kinds of habitats. They are often the dominant autotroph in climates where it's hard for plants to grow, like polar and alpine regions. Lichen diversity is particularly high in temperate rainforests. One place you won't find them, though, is in cities that have significant industrial pollution. When lichens absorb water through their cortex, they also absorb chemical pollutants from the air, making it hard for them to survive in polluted areas. In fact, the U.S. Forest Service monitors and tests lichen populations as part of its air quality program.

FIGURE 3-6: Oakmoss lichen, *Evernia prunastri,* is an example of a foliose lichen.

You can find lichens attached to all kinds of surfaces, both natural and man-made. A few of the places you find them in nature are:

>> **On the bark of trees:** In this case, scientists call them *epiphytes* (*epi*=on, *phyte*=plant).

>> **On bare soil:** Lichens growing on bare soil help form crusts that stabilize the soil. This is particularly important in environments like sandy deserts.

>> **On bare rock:** Scientists call these lichens *saxicolous* lichens (*saxi*=rock, *colous*=living on).

>> **Growing inside rock:** In extremely harsh environments like that of Antarctica, some organisms, including lichens, actually manage to insert themselves under the surface of rocks. *Endolithic* (*endo*=inside, *litho*=rock) lichens grow between grains of minerals in the rock, protected from the elements but still close enough to the surface to get light for photosynthesis.

Lichens live a long time but often grow very slowly. If you were to disturb a lichen crust on the surface of a sand dune, it could take years for the crust to reform. Even though they grow slowly, they play important roles in their ecosystems:

>> **Soil formation:** Lichens can grow on bare rock. As they grow, they help break down the rocks through physical and chemical means, leading to the release of minerals into the soil.

>> **Nitrogen fixation:** Lichens that contain cyanobacteria as photobionts help contribute useful forms of nitrogen to their communities. In some environments where the availability of nitrogen limits plant growth, lichens are an important source of usable nitrogen. The cyanobacterial photobiont captures nitrogen gas (N_2) from the air and incorporates the nitrogen into its own molecules. When bits of the lichen fall to the ground and decompose, the nitrogen becomes available for plants. One environment in which the importance of lichens has been clearly demonstrated is in deserts. Living crusts consisting of different combinations of microbes, including

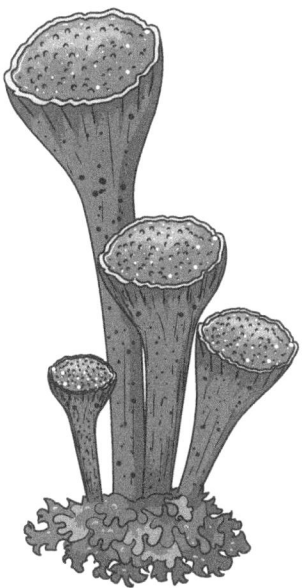

FIGURE 3-7: The pixie cup lichen, *Cladonia asahinae,* is an example of a fruticose lichen.

lichens, cover desert soils. Scientists determined that lichen-containing crusts were able to fix nitrogen when less water was available than were crusts that didn't contain lichens.

» **Food source:** Lichens are autotrophs that make food via photosynthesis. During times of low food availability, organisms can eat some species of lichen. For example, reindeer and people in the far northern areas of Europe and North America eat reindeer lichen, *Cladonia rangifera,* in the winter (see Figure 3-8).

FIGURE 3-8: Reindeer lichen, *Cladonia rangifera.*

Chapter **4**

Examining Fungal Structures

ungi share many similarities with other types of life on Earth, particularly with animals, plants, and other eukaryotes. They also make specialized structures that are unique to their particular lifestyle. In this chapter, you explore the structures that support fungal growth and consider the features that make them unique among living things.

Focusing on Fungi from the Cells Up

As part of Domain Eukarya, fungi are related to animals, plants, and other eukaryotic organisms (see Chapter 2). These similarities are most evident at the cellular level — and cells are the fundamental unit of life on Earth. They can perform all the functions necessary to maintain life and create future generations. Cells can

» **Obtain materials and energy necessary for growth.** Fungi do this by digesting other organisms or food molecules made by other organisms.

» **Use the information in DNA to build cellular structures.** Fungi use the same processes as other eukaryotes to read the code in DNA and use it to build proteins that determine the traits of the organism.

>> **Respond to signals.** Fungi have receptors in their cells that can detect signals like the availability of food and water. They can respond by directing growth toward favorable signals or by turning on cellular defenses in response to negative signals.

>> **Reproduce.** Fungal cells copy their DNA and produce new cells for the next generation. The fungal cell shown in Figure 4-1 is reproducing itself. It has copied all of its components and is building a division, called a *septum*, which will ultimately divide the original cell into two cells.

Examining eukaryotic structures

All eukaryotic cells have similar organization and common structures like those shown in Figure 4-1.

Plasma membrane

The boundary of all cells is defined by a *plasma membrane* (see Figure 4-2), which is a flexible structure built from fat-like molecules called phospholipids and studded with proteins. The plasma membrane separates the inside of the cell from the environment. Although some small molecules, like water, oxygen, and carbon dioxide, can pass directly across the membrane, most materials have to cross with the assistance of a protein, which allows cells to regulate what enters and exits the cell.

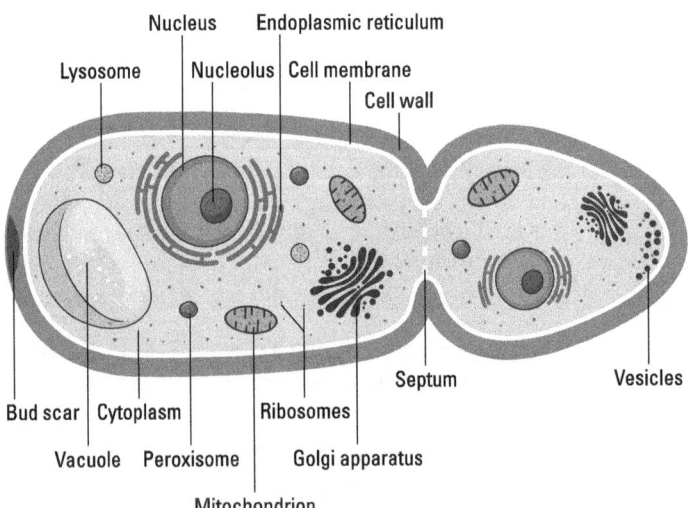

FIGURE 4-1: A fungal cell in the process of cell division.

TIP

You can think of the plasma membrane like an international border where customs agents screen anyone who wants to cross. In this analogy, plasma membrane proteins act as customs agents that screen molecules that want to enter or exit the cell.

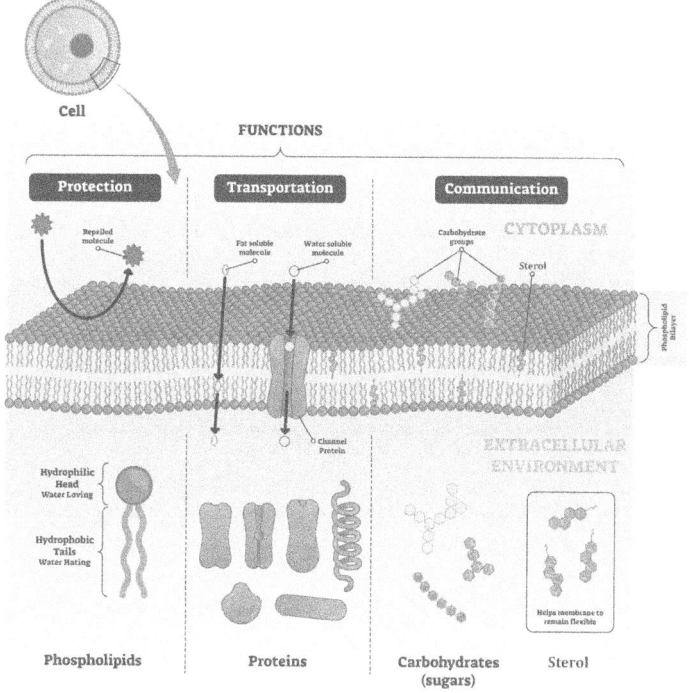

VectorMine/Adobe Stock Photo

FIGURE 4-2: The fluid mosaic model of the plasma membrane.

Plasma membranes also contain small amounts of carbohydrate and sterol molecules. The carbohydrates attach to the outside of the membrane and help in cellular recognition and signaling. The sterol molecules help keep the membrane stable yet flexible. The sterol in animal cell membranes is cholesterol; in fungal cells, it is ergosterol.

Nucleus and DNA

All eukaryotic cells have *organelles*, internal compartments surrounded by membrane borders. The membranes separate the insides of the organelles from the *cytoplasm*, the fluid-filled interior of the cell.

The largest organelle in eukaryotic cells is the *nucleus*, the organelle that stores the DNA. The nucleus is surrounded by a double layer of membrane called the *nuclear membrane*. Like other eukaryotes, fungal cells organize their DNA into *chromosomes*. Each chromosome is a piece of DNA wound around proteins called *histones*. The number of chromosomes in fungi varies between species, anywhere from three chromosomes in some yeast species up to 21 chromosomes in some of the filamentous fungi.

Endomembrane system

The membrane that surrounds the nucleus extends into the cytoplasm of the cell, folding back and forth around the nucleus to form the *endoplasmic reticulum* (ER). The ER is a major manufacturing site in cells:

>> Cells produce some of their proteins on the *rough endoplasmic reticulum (RER)*. The RER got its name because it looks like it is studded with dots when you look at cells through an electron microscope. These dots are *ribosomes,* the cellular structures needed to make proteins (discussed more in the next section). Cells use the RER to make proteins that will either get placed into cellular membranes or be transported out of the cell.

>> Cells produce lipids such as fats, phospholipids, and sterols on the *smooth endoplasmic reticulum* (SER). The SER is called smooth because ribosomes don't attach to it, so you don't see visible dots when you look at the SER using electron microscopes.

A good analogy to help you remember the function of the ER is to think of it as a cellular factory where cells make necessary components.

After cells manufacture molecules at the ER, they ship them to another set of membranes called the *Golgi apparatus*. As you can see in Figure 4-1, the Golgi apparatus looks a little like a stack of pancakes surrounded by small bubbles. The pancakes are stacked compartments surrounded by membranes. The bubbles are *transport vesicles*, small spheres of membrane that cells use to ship materials around.

Cells ship the proteins and lipids made at the ER to the Golgi in transport vesicles. The vesicles fuse with the membranes of the Golgi, just like two soap bubbles can fuse to form one bubble. The proteins and lipids will move through the layers of the Golgi. As they travel, the Golgi makes small chemical changes to the molecules. Some of these changes tag the molecules so that they will be shipped to their proper destination in the cell. When the molecules are ready, they leave the Golgi in another vesicle and travel to their destination.

TIP

A good analogy to help you remember the function of the Golgi apparatus is to think of it as a post office, receiving, tagging, and shipping packages to their proper destinations.

REMEMBER

Cells exchange materials within their internal membrane system, which includes the nuclear membrane, ER, vesicles, Golgi, and plasma membrane. Scientists call the internal membrane system the *endomembrane system*.

Fungi also contain large membrane-enclosed compartments called *vacuoles*. These vacuoles contain digestive enzymes that help fungal cells digest materials that get placed in the vacuole. The vacuole also stores materials and helps maintain the right balance of ions and water in the fungal cell.

Ribosomes

Cells make proteins on cellular structures called *ribosomes*. Ribosomes move around the cell depending upon the final destination of the protein being made:

>> **Free ribosomes** remain in the cytoplasm of the cell. They construct proteins that will remain in the cytoplasm.

>> **Bound ribosomes** travel to the rough endoplasmic reticulum and attach to the membrane. They construct proteins that will either be embedded in a membrane or passed through a membrane to exit the cell.

TIP

You can think of ribosomes as assembly machines in a factory. They assemble a product (proteins) out of materials (amino acids) provided to them by the cell.

Mitochondria

In eukaryotes, *mitochondria* are essential for cellular respiration, the process that enables cells to transfer energy out of food molecules into an energy carrier molecule called *adenosine triphosphate* (ATP). Cells use the energy stored in ATP to do work such as moving materials and building molecules for growth.

REMEMBER

Two membranes surround each mitochondrion, an *outer membrane* and an *inner membrane*. The inner membrane twists back and forth, creating many folds called *cristae*. The space between the two membranes is the *intermembrane space*. The fluid-filled interior of the mitochondrion is the *matrix* of the mitochondrion.

Cellular respiration has three main phases, each of which takes place in a different part of the eukaryotic cell (see Figure 4-3):

>> **Glycolysis:** The breakdown of food molecules, such as the sugar glucose, begins in the cytoplasm of the cell. During glycolysis:

- Cells split glucose in half, forming two molecules of the molecule pyruvate.

- Cells transfer small amounts of energy to the energy carrier ATP.

- Cells also transfer energy and electrons to the electron carrier molecule, *nicotinamide adenine dinucleotide (NADH)*.

>> **Citric acid cycle:** The cell moves pyruvate into the mitochondrion, converting it to a molecule that can enter the citric acid cycle. The citric acid cycle is a series of chemical reactions that takes place in the *matrix* of the mitochondrion. During this cycle,

- Cells continue to break down food molecules, ultimately releasing the carbon atoms from the food back into the environment as carbon dioxide (CO_2).

- Cells transfer a small amount of energy directly to ATP.

- Cells transfer energy and electrons to NADH and another electron carrier called flavin adenine dinucleotide ($FADH_2$).

>> **Oxidative phosphorylation:** The electron carriers NADH and FADH$_2$ release electrons into the *electron transport chain* (ETC), a series of proteins embedded in the inner membrane of the mitochondrion.

- Electrons move through the ETC until they reach oxygen (O$_2$). When the final protein in the chain transfers electrons to oxygen, the oxygen is converted to water (H$_2$O).

- As the electrons move through the ETC, they enable a series of energy transfers that ultimately transfer energy from food into ATP.

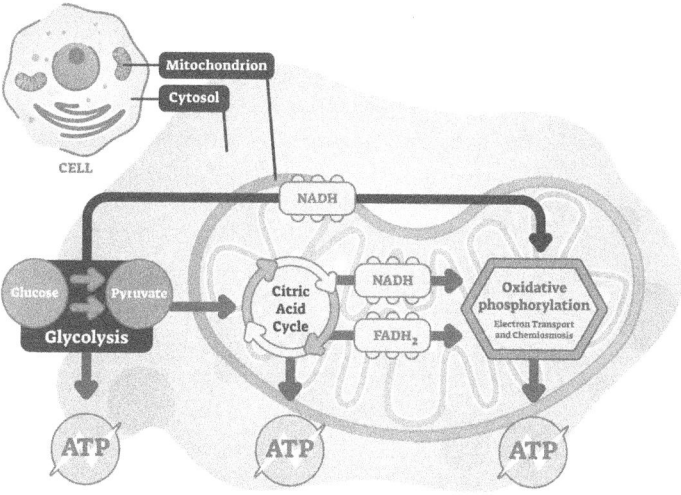

VectorMine/Adobe Stock Photo

FIGURE 4-3: Aerobic cellular respiration in a eukaryotic cell.

TIP

People often call mitochondria the powerhouse of the cell because they know it has something to do with energy. Cells transfer energy from food into a usable form (ATP) in a mitochondrion, just like people transfer energy from coal or moving water to usable electricity in a power plant.

Cytoskeleton

The *cytoskeleton* is a system of proteins that runs throughout cells. These proteins often form long cables that act as tracks for the movement of vesicles and organelles. Cytoskeletal proteins can also act as supportive scaffolding to reinforce cellular membranes. The cytoskeletal proteins *actin* and *tubulin* both play important roles during fungal growth and cell division:

» The protein *actin* helps determine the direction of growth of fungal cells:

- Filamentous fungi require cables of *filamentous actin* (F-actin) in order to grow from the tips of their hyphae. The F-actin provides tracks for the delivery of vesicles containing enzymes and new material for the cell wall at the growing tip.

- In fungi growing as yeast, *actin patches* and F-actin gather and orient themselves at the site for the formation of the new cell. Some of the actin migrates into the new cell as it grows off of the original, delivering the enzymes and wall materials for the growing cell.

- When a growing fungal cell is ready to separate into two cells, an *actin ring* forms at the location where the new wall will form. With the help of the protein *myosin*, the actin ring contracts as the new wall is put down, forming a cross-wall, or *septum,* between the two cells.

» The protein tubulin forms long filamentous cables called *microtubules*:

- During cell division, microtubules form the *mitotic spindle,* protein cables that sort the chromosomes properly into the new cells.

- During cell growth, microtubules serve as tracks to help distribute organelles and vesicles throughout the cell.

Two good analogies for remembering the function of the cytoskeleton are to think of it as railroad tracks and scaffolding.

TIP

Finding out about unique features

Although the fundamental cell structure of fungi is similar to that of other eukaryotes, fungal cells do have some unique characteristics. Because some of these structures are found only in fungal cells, they represent potential targets for antifungal drugs and the immune systems of organisms infected by fungi.

Cell wall

Like plants and bacteria, fungi have an additional layer outside their plasma membranes called a *cell wall*, shown in Figure 4-4. The fungal cell wall has similar functions to the cell walls found in other organisms:

>> Cell walls are rigid and protect the cell from bursting due to changes in water pressure, which is called *osmotic lysis*.

>> Cell walls have molecules on their surface called *adhesins* that help fungal cells attach to other cells or surfaces.

Although the function of fungal cell walls is similar to that of other cells, the chemical composition is unique. Although the exact composition varies between species of fungi, most fungal cell walls have certain things in common:

>> Fungal cells deposit the nitrogen-containing polysaccharide *chitin* just outside the plasma membrane of the cell. Chitin is a tough polysaccharide that's also found in the exoskeletons of insects. In fungi, it can be anywhere from one to 20 percent of the cell wall.

>> Outside of the chitin layer is the most abundant component of the wall, structural polysaccharides called *glucans*. These carbohydrates can make up 50 to 60 percent of the cell wall.

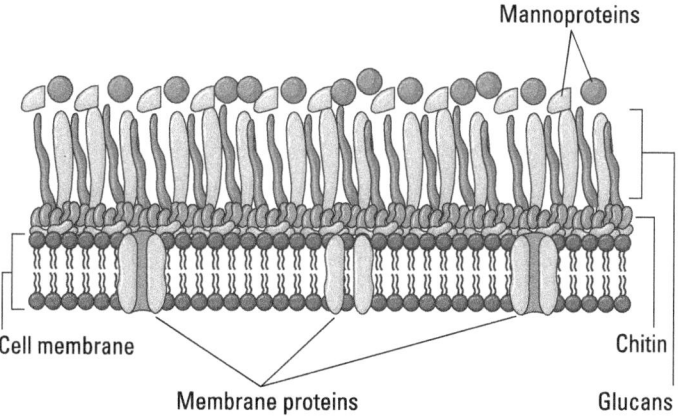

Mannoproteins

Cell membrane

Membrane proteins

Chitin

Glucans

FIGURE 4-4: The structure of the fungal cell wall.

>> *Glycoproteins,* which are proteins linked to sugars, form the outermost layer of the fungal cell wall and can be anywhere from 20 to 50 percent of the cell wall. One example of these types of proteins is the mannoproteins shown in Figure 4-4.

Chitosomes and microbodies

Chitosomes are small vesicles that transport the enzyme *chitin synthetase* to the parts of cells that are actively growing. The vesicles deliver the enzyme so that it can speed up the synthesis of chitin needed for the formation of new cell walls. Chitosomes are also called microvesicles because of their small size. The larger vesicles that carry the building blocks for chitin to the site of growth are called macrovesicles.

The name of the enzyme *chitin synthetase* tells you what it does. It is an enzyme (name ends in -*ase*), so it speeds up a reaction. In this case, it speeds up the reaction that synthesizes the polysaccharide chitin.

Microbodies are small organelles formed from a single membrane surrounding a protein-filled fluid. These organelles participate in various aspects of fungal metabolism. Some contain enzymes for breaking down fatty acids, others have enzymes for digestion of unusual carbon and nitrogen sources. Scientists give some types of microbodies names that reflect a specific function. For example, peroxisomes are microbodies that contain the enzyme peroxidase. Glyoxysomes convert stored fats to sugars through a pathway called the glyoxylate cycle.

Starting with Spores

If you've ever seen little black dots on the fuzzy mold growing on your bread or powdery green dust on some cheese mold, you've seen fungal spores, like the ones shown in Figure 4-5.

Spores are reproductive cells that are specialized for dispersal or survival.

Fungi can make spores as part of either asexual or sexual reproduction (see Figure 4-6):

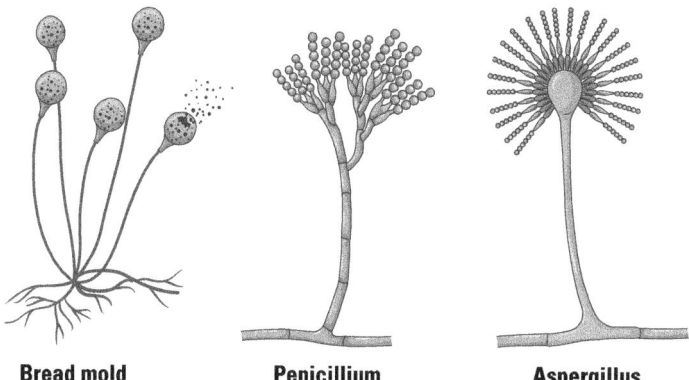

Bread mold   **Penicillium**   **Aspergillus**

FIGURE 4-5: Asexual spore formation in three types of mold.

» **Asexual reproduction** occurs when an individual organism reproduces itself without genetic input from another individual. Fungi can produce spores asexually using a type of cell division called *mitosis,* which makes exact copies of the nucleus of a cell.

- Some fungi make asexual spores called *conidia* at the tips of their hyphae, forming little broom-like structures, which you can see on the mold in the center and on the right of Figure 4-5. The hyphae that hold up the conidia are called *conidiophores* (*-phore* means *to carry*).

- Other fungi make asexual spores inside little sacs called *sporangia,* like the ones shown on the mold on the left in Figure 4-5. These spores are sometimes called *sporangio-spores.* The hypha that holds up the sporangium is called a *sporangiophore.*

» **Sexual reproduction** involves the fusion of two nuclei to create a new individual that has genetic material from two different cells. Although different groups of fungi vary in the details of sexual reproduction (presented in Part 2 of this book), the process typically occurs in three steps:

- Two hyphae fuse together, combining the cytoplasm and nuclei of two cells into one. The fusion of the cytoplasm is called *plasmogamy.*

- The two nuclei fuse together, combining their genetic information. The fusion of the nuclei, called *karyogamy,* temporarily doubles the amount of genetic information in the nucleus of the cell.

CHAPTER 4 **Examining Fungal Structures**   63

- This cell divides by a special type of cell division called *meiosis,* which reduces the genetic information in half, returning it to the original amount. The resulting cells, which may also divide by mitosis to increase their number, become sexual spores. The spores produced by this sexual process contain a mix of genetic information from the two nuclei.

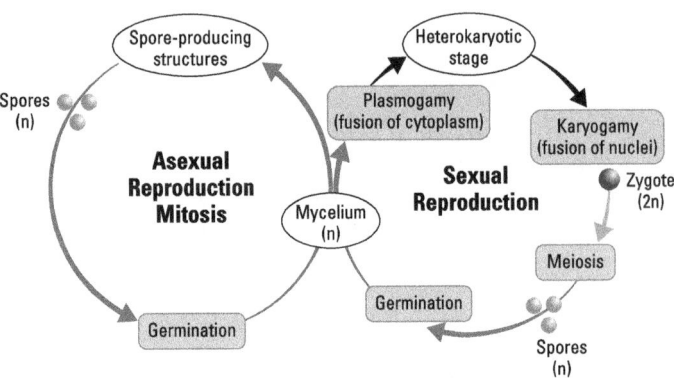

FIGURE 4-6: An overview of reproduction in fungi.

Some groups of fungi form unique structures during sexual reproduction. Figure 4-7 shows three examples of these structures. The Zygomycetous fungi (Chapter 9) form a thick-walled *zygosporangium* that contains a single zygospore. The Ascomycota (Chapter 10) form their sexual spores within a small sac called an *ascus.* The Basidiomycota (Chapter 11) form their spores on club-shaped structures called *basidia.*

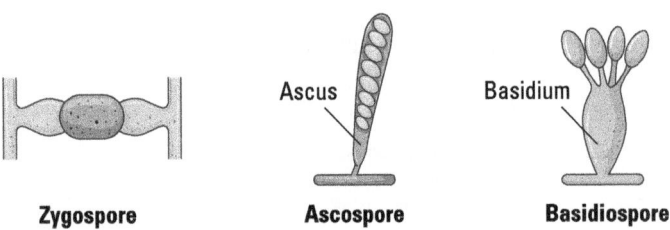

FIGURE 4-7: Sexual spores in three groups of fungi.

In gilled mushrooms, the basidia form on the gills located under the mushroom cap. To see the spores and their color, you can cut off the stem of the mushroom and place the mushroom cap on a piece of paper or glass, leaving it undisturbed for several hours (anywhere from 2 to 24 hours; less time is needed for fresh mushrooms). It can help to put a drop of water on top of the mushroom cap to help release the spores and to cover your mushroom print with a cup or bowl so air currents don't blow your spores around. When you are ready, carefully lift your mushroom cap to reveal a spore print. To preserve your spore print, you can spray it with acrylic spray or even hairspray. To avoid disturbing your spores, spray the back side of the paper and let the adhesive soak through.

Spreading spores

Fungal spores are small and very light, so they literally blow away on a breeze to new locations. Fungi don't just rely on random wind, however. Instead, they have several extremely fun methods of making sure their spores go places:

>> The dung fungus *Pilobolus* performs one of the most famous methods for spore dispersal so fast that it's sometimes called a fungus cannon. The fungus produces a single black sporangium at the tip of a sporangiophore. A vesicle just below the tip of this hypha swells with water, creating a great deal of pressure. The fungus has receptors that can detect light, so the sporangiophore rotates and keeps the sporangium pointed toward the sky.

When the fungus is ready to disperse its spores, the swollen vesicle blows apart, throwing the black sporangium into the air. The acceleration achieved by this explosive release is actually faster than that of a bullet from a gun. The sporangium is flung up to two meters away, looking like a little black hat being flung off the swollen head of the hyphal tip (which gives this fungus the nickname of the hat-throwing fungus). The fungal spore can begin to grow in a new area of grass, where the fungus will eventually be eaten by a grazing animal, eventually producing spores in a new batch of dung.

>> Puffballs are another group of fungi that give their spores a little extra boost to catch the wind. Puffballs are spherical fungi that grow to different sizes depending upon the species, from about the size of a golf ball to as large as a beach ball (see Chapter 16 for more on the giant puffball).

Puffballs produce millions of spores along the inside edges of cavities within the ball. They are edible when they are young and will appear fleshy and white if you slice them open. As they age and the spores mature, the outer layer of the puffball, called the exoperidium, begins to wear away. At this point, all it takes is a small amount of pressure from a raindrop or a passing animal to cause of the puffball to release its spores. Some puffballs have a small opening, called an ostiole, that becomes visible when the exoperidium is gone. Massive amounts of spores can be channeled through this opening like a plume of smoke. People have been joyously stomping on puffballs to make them smoke, probably as long as there have been people.

>> Some fungi get help from other species to disperse their spores. The fungus that causes Dutch elm disease is carried from tree to tree by bark beetles. The beetle larvae feed on the inner bark of the tree. When they emerge as adults, sticky fungal spores cling to their backs. The beetles travel to new trees, taking the spores with them. As they chew through the bark in order to lay their eggs, the fungal spores get scraped off onto the tree.

Scientists have also found fungal spores on the bodies of other insects, such as ground beetles. As insects move between areas, they may carry fungal spores with them, helping to move fungi within ecosystems.

>> Some species of aquatic fungi (Chapter 8) produce spores that can swim. These zoospores produce cell structures called *flagella* that can flex back and forth because of their internal cytoskeletal proteins.

Germinating when the time is right

Most spores will remain dormant until the growing conditions are right. Once they encounter the right temperature, moisture, and food availability, they will *germinate* or begin to grow, as shown in Figure 4-8. During germination, several things happen:

>> **Hydration:** Water enters the spore.

>> **Increased metabolism:** The rate of cellular respiration increases as the fungus processes food molecules more rapidly.

>> **Increased biosynthesis:** The fungus starts to manufacture the molecules it needs to build more cells.

>> **Outgrowth:** A new hypha emerges from the spore, as shown in Figure 4-8. Fungi of the same species may produce hyphae that have specific mating types (shown as "+" and "–" in Figure 4-8). The mating type of hyphae can determine whether two individuals are sexually compatible.

Some fungal spores require an environmental shock to trigger germination. The spores of fungi that grow on dung have to pass through the digestive tract of an animal before they begin to grow. Other fungal spores have to pass through the heat of a fire before they will germinate.

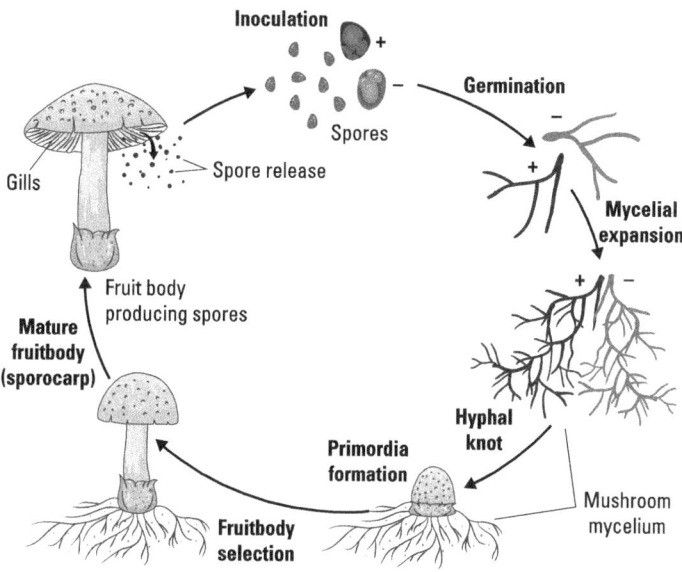

FIGURE 4-8: The life cycle of a mushroom.

Weaving Through the World

Fungi hyphae grow in length as cells divide to add new cells to the growing chain. The chains can branch, sending filaments in all directions throughout the substrate that the fungus is using for food, as shown earlier in Figure 4-8.

Growing from the tip

REMEMBER

Fungal growth occurs at the tips of hyphae. The materials and cellular structures needed for growth organize in a structure known as the *Spitzenkörper*, which is a German word meaning pointed body.

The Spitzenkörper is located just behind the tips of actively growing hyphae, including the hyphae emerging from a germinating spore or those forming a branch from an existing hypha. The Spitzenkörper consists of vesicles and chitosomes (see the close-up in Figure 4-9) within a web of filamentous actin. The vesicles contain the building materials needed for new wall construction, and the chitosomes contain chitin synthetase.

Because the vesicles found in the Spitzenkörper come from the endoplasmic reticulum and pass through the Golgi apparatus, the Spitzenkorper is part of the endomembrane system of the fungal cell.

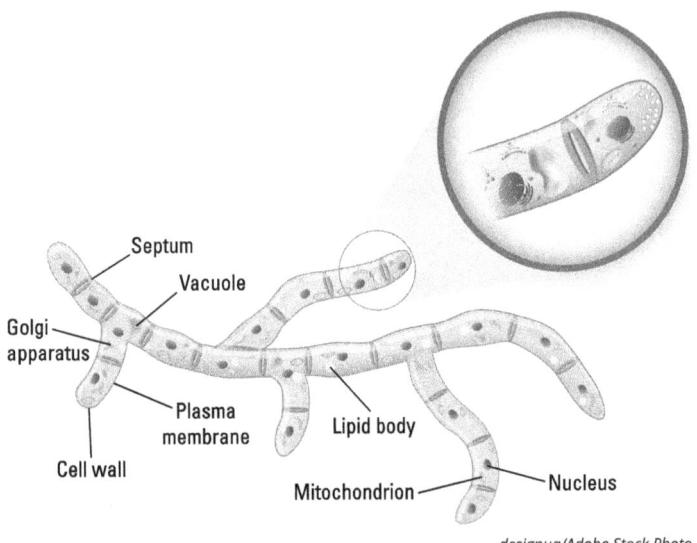

designua/Adobe Stock Photo

FIGURE 4-9: Fungal hypha with a close-up of the hyphal tip.

Most hyphae have regular cross walls that separate the cytoplasm into separate compartments, like the hypha shown on the top in Figure 4-10. Some fungi form very few cross walls, like the hypha shown on the bottom in Figure 4-10.

The cross-walls between fungal cells are called *septa* (singular=*septum*). Fungi with cross-walls that divide their hyphae have *septate hyphae*. Hyphae that lack cross-walls are *coenocytic*.

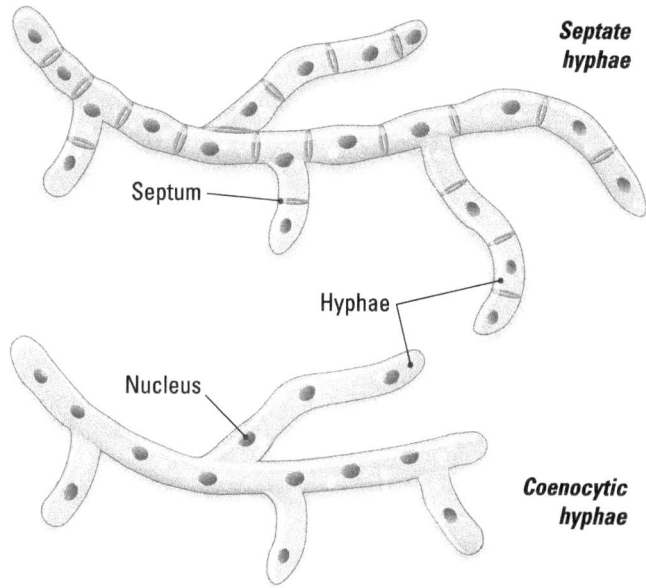

FIGURE 4-10: Septate and coenocytic hyphae.

Forming mycelia

Fungal hyphae continue to grow and branch, forming an interwoven web of hyphae called a *mycelium,* like the one shown earlier in Figure 4-8. The body, or *thallus,* of a fungus is made from this mycelium. Some fungi grow as microscopic mycelia for most of their life cycles, while others periodically organize their mycelia into complex structures.

Recognizing Different Species

If you are someone who likes to know what you're looking at when you're exploring nature, you may want to identify some of the fungi you encounter. The good news is that some fungi are fairly easy to recognize; the bad news is that some can only be identified using a microscope. A guide to your local fungi with good pictures can be very helpful. If you want to identify fungi so that you can forage and eat them, my advice is to work with someone who has foraging experience in order to learn which ones are safe.

TIP

One great place to find experienced foragers who know the mushrooms in your area is in a local mushroom club. You can search the Internet for a club to join. A good place to start is the North American Mycological Society (NAMA), which lists affiliated clubs on its website at https://namyco.org/clubs.

People typically identify fungi by the appearance of their reproductive structures, which are called *sporocarps* or *fruiting bodies*. Characteristics that can help you identify a fungus are:

>> **Shape and size:** Fruiting bodies may be a wide variety of shapes, including spherical, shelf-like, umbrella-shaped, star-like, and downright phallic. As fruiting bodies grow, their shapes can change. For example, umbrella-shaped mushrooms start as small buttons or balls before they elongate their stipes and then spread their caps. Mature fungal fruiting bodies range in size from barely visible to several feet in diameter.

>> **Color and texture:** Some fruiting bodies are firm, others are jelly-like, and some are dry and brittle.

>> **Smell:** Some fruiting bodies smell spicy, fruity, or woody, while others smell like urine or rotting cabbage.

>> **Location:** Some fungi grow in certain environments or in association with particular plants.

Mapping a mushroom

One of the reasons fungi are so popular is that many of them produce absolutely gorgeous fruiting bodies. Their colors range from bright yellow, white, and brown to bright red and even lavender. They grow in a wide variety of shapes, from skinny stems to bulbous and from flat to pointed caps.

REMEMBER

When we think of mushrooms, we often think of the traditional shape, which consists of a stem called a *stipe* and an umbrella-like *cap*, like those shown on the mushroom in Figure 4-11. The technical term for a cap is *pileus*.

The reproductive structures that form sexual spores usually develop on a surface called the *hymenophore*. The location of the hymenophore varies among fungal species. Possibilities include:

>> **Gills:** Many mushrooms have *gills* (shown in Figure 4-11), which are thin blades of tissue that radiate around the cap. Several features of gills can be helpful for identification:

- *Spacing of gills.* The distance between individual gills differs between some species of fungi.

- *The color of the gills.* The gills may be a different color from the cap. If you bruise the gills by pressing on them with something, the color may change. Bruising can also release a milky substance in some fungi.

- *Forked gills.* Gills may travel in straight lines from the edge of the cap to the stipe, or they may fork and branch.

- *How the gills attach to the stipe.* In some fungi, like oyster mushrooms, the gills run down the stem of the mushroom. In others, like portobello mushrooms, the gills don't attach to the stipe at all.

>> **Pores:** Some mushrooms have little holes called *pores* on the underside of the cap. These pores are actually the open ends of tubes that line the underside of the fruiting body. Color, pore size, and pore pattern can help you identify these fungi. Like gills, pores may change color when bruised. Porcini mushrooms are a type of bolete mushroom that have pores on the underside of their caps.

>> **Teeth:** Spores can also be produced on long dangling extensions called *teeth* or *spines*. The teeth hang down from the underside of the cap and can give the mushroom a shaggy appearance, as in the lion's mane mushroom.

WARNING

Some fungi have *ridges* that resemble gills. True gills look like thin blades beneath the flesh of the cap. Ridges are wrinkles that are part of the cap and can't be separated from it.

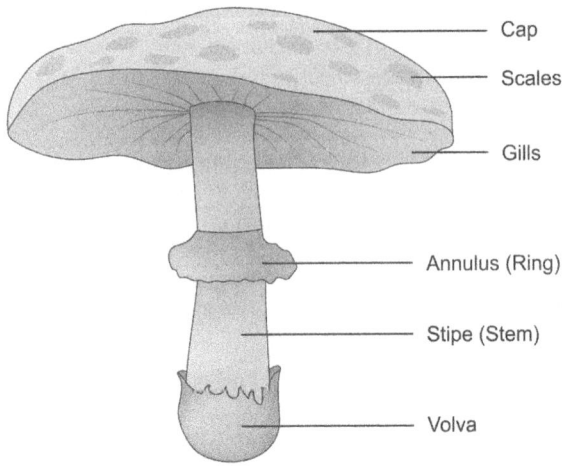

Cap

Scales

Gills

Annulus (Ring)

Stipe (Stem)

Volva

FIGURE 4-11: Mushroom anatomy.

Mushrooms may also have distinctive features associated with a veil. A *veil* or *velum* is a thin layer of tissue that covers some mushrooms as they develop. The veil helps protect the developing spores by maintaining the right amount of moisture. As the mushrooms grow and expand, they tear their veils. The remnants of the veils can sometimes be seen on the mushroom:

>> **Universal veils** completely cover mushrooms as they begin to grow. These veils may leave several types of remnants:

- **Scales or warts:** Scales, like those shown in Figure 4-11, are flaky or shaggy patches that are typically scattered across the cap but may also appear on the stipe. Warts are bumps on the cap.

- **A *volva*:** The volva is a cup-like structure at the base of the stipe formed from the remains of the bottom of the veil (see Figure 4-11).

- **Rings:** The veil may also leave behind a ring, or *annulus* (see Figure 4-11), around the stipe. Sometimes multiple rings may be seen.

- **Powdery deposits:** These are fine particles left behind on the stipe or cap.

>> **Partial veils** extend from the edge of the cap toward the stipe, creating a protective barrier under the cap. Partial veils

may leave a ring of tissue around the stipe. This tissue may be firm or hang down in a thin sheet like a skirt.

Paying attention to place

Fungi form intimate connections with their environment. Knowing where a fungus likes to grow can help you find and identify it. When hunting for particular mushrooms, here are some things to think about:

>> **Substrate:** Noticing the specific habitat of a fungus can help you narrow your identification. Some fungi grow on rotten wood or leaves; others may grow in a grassy field. Some can grow on the trunks of live trees.

>> **Nearby trees:** Fungi that form partnerships with particular trees will be found growing near that species. Learning to recognize these trees can help you locate these fungi.

>> **Season:** Some fungi form their fruiting bodies during a specific season. If you are looking for a particular type, it's important to know when to look. Seasonal changes can trigger reproduction, so it's important to keep an eye on the temperature.

Chapter **5**

Reproducing the Fungal Way

Fungi move through their life cycles in a variety of different ways. Most species can reproduce sexually, but many fungi grow and reproduce asexually most of the time. Asexual reproduction usually produces offspring that are genetically identical to the parent, although some fungi have strategies for creating genetic variation during this process. If an organism is highly successful in its current environment, then its genetically identical offspring will also probably be successful. And, as a bonus, asexual reproduction doesn't require that an organism find a compatible partner.

On the other hand, sexual reproduction has the advantage of producing offspring that are different from the parents. If an environment is changing or is stressful in some way, producing offspring with new combinations of traits may ensure that some of them survive. In this chapter, you get an overview of different fungal life cycles and take a look at the details of how fungi reproduce.

Exploring Fungal Reproduction

Fungi begin their lives as spores, and they reproduce by making spores, either through sexual or asexual processes (Chapter 4). As they go through their life cycles, some fungi grow almost invisibly by hyphae and only catch most people's eye when they produce a larger sexual structure like a mushroom or when they show up as fuzzy, colorful patches on spoiled food. Other fungi produce two different visible forms as part of their life cycles, one during sexual reproduction and another during asexual reproduction. In some cases, these two forms are so different that they weren't recognized as belonging to the same species until scientists developed the ability to read the genetic code.

For fungi that make more than one form, scientists use the following terms:

>> The *teleomorph* is the structure made during sexual reproduction, such as a mushroom or a bracket fungus.

>> The *anamorph* is the structure made during asexual reproduction. Anamorphs often look mold-like.

>> The *holomorph* refers to the whole fungus, including both forms.

For fungi that were assigned two names, but later discovered to be one fungus, the name of the teleomorph became the scientifically accepted name. Although anamorph names aren't considered scientifically valid anymore, some people still use them.

Like other organisms that undergo sexual reproduction, the number of chromosomes found within a fungal cell can change depending on its stage in the life cycle. When two individuals combine their genetic information, they produce cells that have two copies of every chromosome.

Cells that have two sets of chromosomes are *diploid*. Cells that have one complete set of chromosomes are *haploid*.

Thinking about human cells may help you wrap your brain around the idea of diploid and haploid cells. Human body cells have 46 chromosomes, but human eggs and sperm only have 23 chromosomes. Body cells are diploid, while eggs and sperm are haploid. To create eggs and sperm, body cells in the ovaries and testes undergo a special type of cell division called *meiosis* that cuts the chromosome number in half, from 46 to 23. This way, the fusion of egg and sperm creates a new cell that again has 46 chromosomes. Another way to think about it is that human cells have 23 types of chromosomes. When an egg and sperm combine their nuclei, each gives one of each of the 23 types, which is why body cells have 46 chromosomes (23 pairs, one of each from each biological parent).

Scientists represent one set of chromosomes as the letter "n." A haploid cell has n chromosomes, whereas a diploid cell has 2n chromosomes.

In humans, n=23. Haploid human cells (sperm and egg) have 23 chromosomes. Diploid human cells (body cells) have 46. Chromosome numbers in fungi are presented in Chapter 6.

As fungi grow and go through their life cycle, they reproduce their cells using two types of cell division:

>> *Mitosis* makes exact copies of cells and does not normally change the chromosome number.

>> *Meiosis* produces cells that have half the genetic information of the parent cell; in other words, it reduces the chromosome number from 2n to n.

Figure 5-1 shows how mitosis and meiosis play a role in fungal life cycles:

>> **Budding yeast (see Figure 5-1A), like the baker's yeast *Saccharomyces cerevisiae*, typically grow as single cells.** Haploid cells can continue to grow indefinitely, producing new haploid cells by mitosis. If haploid yeast cells with opposite mating types encounter each other, they can merge, combining their nuclei to produce diploid cells. These diploid cells can continue to grow asexually, producing more diploid cells by mitosis. If environmental conditions become stressful, then the diploid cells may enter meiosis, producing four haploid cells within a sac called an *ascus*.

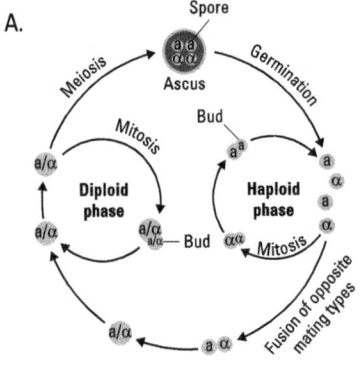

Budding yeast

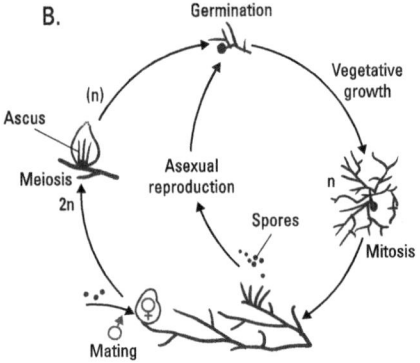

Filamentous ascomycetes

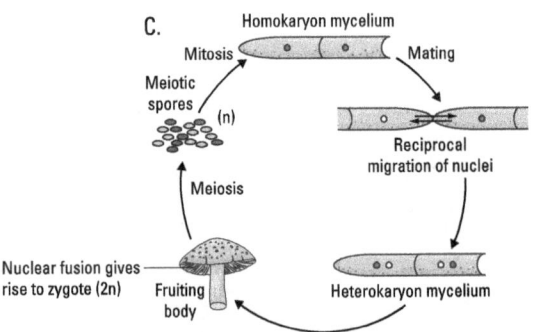

Filamentous basidiomycetes

FIGURE 5-1: Fungal life cycles.

>> **Filamentous ascomycetes (see Figure 5-1B) germinate from haploid spores and grow into hyphae by using mitosis to produce chains of haploid cells.** They typically continue to grow and reproduce asexually, forming asexual spores called *conidia* (or conidiospores) by mitosis (for more on conidia, head back to Chapter 4). If individuals of compatible mating types encounter each other, they may join their hyphae together (plasmogamy; see Chapter 4) so that a nucleus from each individual is in the same cell. Then the nuclei may fuse together (karyogamy) to produce a diploid cell. Meiosis occurs soon afterwards, producing haploid cells that form an ascus and haploid spores.

>> **Filamentous basidiomycetes (see Figure 5-1C) germinate from haploid spores, producing hyphae by mitosis that have a single type of nuclei (homokaryon, *homo*=same, *karyon*=seed).** When two individuals with compatible mating types encounter each other, their hyphae fuse so that they combine cytoplasm (plasmogamy) and exchange nuclei, leading to the creation of cells that have one nucleus from each individual. This cell, called a *dikaryon* (*di*=two, *karyon*=seed), continues to grow with both nuclei dividing synchronously by mitosis. Environmental signals can trigger the fungus to form a fruiting body. Within a cell of the fruiting body, the two nuclei finally fuse (karyogamy), and meiosis follows immediately to produce haploid spores.

Most fungi spend the majority of their life cycle in the haploid state.

REMEMBER When two fungal nuclei combine their genetic information, meiosis usually follows soon after to return the fungus to the haploid condition.

Mitosis

Cells divide by mitosis to produce exact copies of the parent cell. Fungi use mitosis to grow and reproduce. For example,

>> Yeast divide by mitosis to reproduce asexually.

>> In filamentous fungi, mitosis enables hyphae to grow longer by adding cells at the tip of the hypha. Hyphae can also use mitosis to produce asexual spores called conidia.

Mitosis is just one stage in a series of events that scientists call the *cell cycle* (see Figure 5-2). Scientists divide the cell cycle into two main parts:

>> **Interphase occurs when a cell isn't dividing.** Interphase can be further broken down into three phases:

  ● **Cells grow in G1, which stands for "gap one."** Cells can stay in G1. If the cell detects a signal to divide, it will move into S phase.

  ● **Cells get ready for cell division by copying their DNA during S phase, which stands for "synthesis phase."** After they copy their DNA, they move into G2.

  ● **Cells check their DNA in G2, which stands for "gap two."** When they are ready, they proceed into M phase.

>> **Cells divide during M phase.** During growth or asexual reproduction, M stands for mitosis. During sexual reproduction, M represents meiosis.

To produce exact copies via mitosis, cells need to be extremely organized about duplicating their chromosomes and then separating them into the new cells.

>> During S phase, cells use each chromosome as a pattern to build an exact copy of the original. The two copies stay attached together by a region called the centromere (more on these in Chapter 6). Each copy is called a *sister chromatid,* and the entire duplicated chromosome is a *replicated chromosome* (see Figure 5-2).

>> A special organelle, called the *mitotic spindle,* attaches to the replicated chromosomes during mitosis and guides one of each sister chromatid to opposite cells. This makes sure that each new cell has a complete set of chromosomes.

For cells to move their chromosomes around efficiently, they first need to package them into tight bundles. During interphase, chromosomes are relatively loosely wound around proteins so that the DNA spreads out through the nucleus. Scientists refer to this relaxed form of the chromosomes as *chromatin*. If you were to use a stain that sticks to DNA and then view a cell in interphase under the microscope, the entire nucleus would show the color. You wouldn't see distinct bundles of each chromosome. Cells coil their chromosomes more tightly during the early part of mitosis so that each becomes a distinct structure.

TIP

If you think of DNA like cold pasta, chromatin would be a tangle of long spaghetti. Coiled chromosomes would be more like short spiral noodles, such as rotini. If you had to sort each piece of pasta away from the others, it would be much easier to sort the short noodles.

Another thing cells need to do to get ready for mitosis is to build their mitotic spindle. The spindle forms from a *microtubule organizing center* (MTOC) called the *spindle pole body.* During S phase, cells copy their spindle pole body so they have two. The two spindle pole bodies move away from each other and take up positions on opposite sides of the cell, as shown in Figure 5-2.

REMEMBER

The *mitotic spindle* is part of the *cytoskeleton* of the cell. Long, cable-like proteins called *microtubules* grow outward from two *spindle pole bodies,* forming a network of *spindle fibers* like those shown in Figure 5-2. Some spindle fibers attach to proteins called *kinetochores,* which are positioned at the centromeres of replicated chromosomes. During the later stages of mitosis, the spindle pole bodies disassemble the microtubules, essentially reeling the spindle fibers back in. As the microtubules shorten, they pull the two sister chromatids from each replicated chromosome into separate cells, as shown in Figure 5-2.

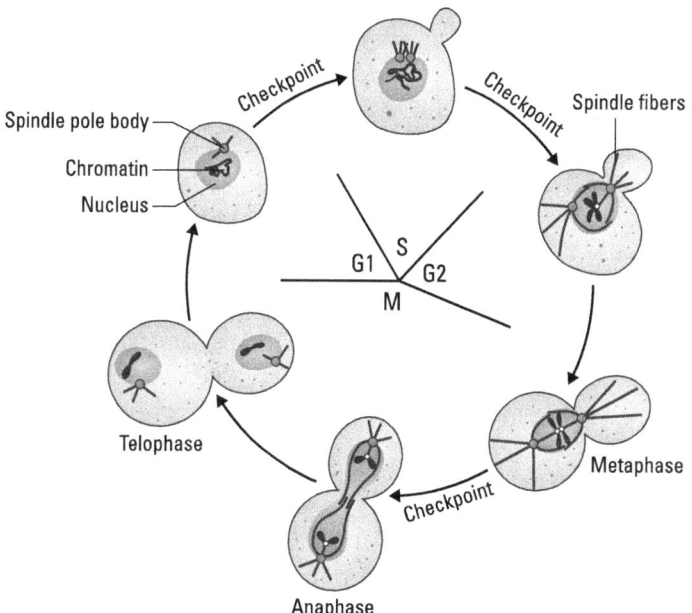

FIGURE 5-2: The cell cycle in yeast.

The major events of mitosis are the same in fungi as they are in other eukaryotes, such as plants and animals, but there are some unique features of fungal mitosis. Scientists divide the events of mitosis into phases based on the events that take place in each phase:

>> **During *prophase*, chromosomes coil up (condense) and spindle fibers attach to chromosomes.** In plants and animals, the nuclear membrane breaks down, but many fungi have *closed mitosis* during which the nucleus remains intact.

>> **During *metaphase*, the spindle fibers push and pull the chromosomes to organize them prior to separation.** In plants and animals, cells line the chromosomes up in the middle of the cell at the halfway point between the two poles of the spindle. In fungi, the chromosomes spread out a bit more, occupying the middle third to half of the spindle region.

>> **During *anaphase*, the pairs of sister chromatids release from each other, and the mitotic spindle begins to shorten, pulling the chromatids to opposite poles.** In fungi, anaphase happens in two steps:

- *Anaphase A* occurs when sister chromatids release from each other.

- *Anaphase B* begins when sister chromatids reach the poles. At this point, the spindle elongates rapidly, moving the two pools of DNA farther away from each other.

>> **In animals and plants, *telophase* refers to the reconstruction of a distinct nucleus.** The nuclear membrane reforms, and chromosomes uncoil, returning to their relaxed state as chromatin. In most fungi, the nuclear membrane never breaks down. Telophase usually occurs when cytoskeletal proteins constrict like a belt around the nuclear membrane, causing it to collapse and form two nuclear membranes, one around each set of chromosomes.

REMEMBER

Mitosis refers to the division of the nucleus. *Cytokinesis* refers to the division of the cytoplasm.

As mitosis finishes, many cells separate their cytoplasm so that two completely new cells result. In plants and animals, cytokinesis often occurs at the same time as telophase. This is true for some fungi, such as yeast, and is true for most fungi during spore formation. However, for filamentous fungi, cytokinesis doesn't always follow mitosis. Instead, some hyphae continue to grow longer without forming cross-walls between new nuclei. These *coenocytic hyphae* (Chapter 4) can contain multiple nuclei.

REMEMBER

Cytokinesis in fungi is very similar to that in animal cells. A band of cytoskeletal proteins containing actin and myosin (the same proteins in animal muscles) contracts and pinches the plasma membrane into a furrow. This furrow forms around a new wall septum that grows out from the existing wall.

TECHNICAL STUFF

Some of the events of mitosis occur early in budding yeast. In most eukaryotes, the mitotic spindle doesn't begin to form until mitosis begins, but yeast form short spindle fibers in S phase. Likewise, cytoskeletal proteins make a ring and begin to pinch the plasma membrane to form a bud in late S phase.

As cells grow, they respond to internal and external signals to determine if they should divide. For example, cells monitor their environment for conditions such as the temperature, availability of food and water, and the amount of daylight. If conditions are right, the cell might make proteins that signal the cell to enter cell division. Before a cell moves forward in the cell cycle, however, it stops and checks to make sure it's ready. *Checkpoints* are the key moments a cell commits to moving from one phase of the cell cycle to another. Three checkpoints that are particularly important for controlling cell growth are shown in Figure 5-2:

>> **G1 checkpoint:** Once a cell moves out of G1 and into S phase, it's committed to cell division. So, before it progresses, it checks to make sure that it's big enough and has enough energy. It also checks to make sure its DNA isn't damaged. If the cell doesn't pass this checkpoint, it can remain in G1 or move to a resting phase called G0 ("G zero").

>> **G2 checkpoint:** After cells copy their DNA in S phase, they pause in G2 to make sure all the chromosomes are copied and to check for any breaks in the DNA. If the cell detects problems, it can pause in G2 and try to repair itself.

>> **M checkpoint:** During metaphase of mitosis, the cell checks to make sure the mitotic spindle is attached to all of the chromosomes. If it finds any unattached chromosomes, it will pause for the spindle to pick them up.

The cell cycle and its quality control system of checkpoints are very similar among eukaryotes. Mistakes in the process do slip through. In multicellular eukaryotes, these mistakes may result in programmed cell death to remove defective cells. If the cells don't die, they may contribute to syndromes or the development of cancer. In unicellular organisms, mistakes may lead to cell and organism death, or they may be tolerated. In the case of many fungi, mistakes that lead to chromosomal abnormalities may be advantageous, particularly during environmental stress.

Meiosis

Cells divide by meiosis to produce cells with half the genetic information of the parent cell.

Meiosis reduces the number of chromosomes from diploid to haploid.

REMEMBER

Cells that divide by meiosis follow the same phases of the cell cycle as those that divide by mitosis (shown earlier in Figure 5-2). Under the right circumstances, however, the cell will enter meiosis instead of mitosis. For fungi, the conditions that trigger meiosis often involve the presence of a compatible mating partner and environmental stress, such as a lack of food. Under these conditions, a fungus may switch from asexual reproduction to sexual reproduction. In the case of filamentous fungi, two haploid nuclei will fuse together to create a diploid nucleus. The diploid nucleus then divides by meiosis to produce haploid spores that will begin the next generation.

One evolutionary advantage of sexual reproduction is that it creates offspring that are different from the parents. If environmental conditions are stressful, a new combination of genes may favor survival.

REMEMBER

In addition to following the same steps of interphase (G1, S, G2), the phases of meiosis have similar names and main events as the phases of mitosis. One important difference, however, is that meiosis has two rounds of division, called meiosis I and meiosis II.

Cells go through S phase prior to meiosis, copying their chromosomes to form replicated chromosomes. To prepare for sexual reproduction, cells need to divide the replicated chromosomes twice, as shown in Figure 5-3B:

>> During *meiosis* I, cells separate the matching pairs of chromosomes to create haploid cells that have replicated chromosomes.

>> During *meiosis II*, cells separate the sister chromatids of the replicated chromosomes.

TECHNICAL STUFF

Cells use the mitotic spindle to carefully organize the process of meiosis. Scientists divide the two rounds of division into phases based on the movement of the chromosomes. The major events of these phases are very similar to those of mitosis, but there are some key differences:

>> **During meiosis I,** matching pairs of replicated chromosomes separate (see Figure 5-3B).

   ● **During prophase I,** the matching pairs of chromosomes, called *homologous chromosomes,* pair up and stick together. Remember that diploid cells have two complete sets of chromosomes, meaning they have two of each unique chromosome for their species. (In humans, this would be two copies of 23 chromosomes for a total of 46.) When cells divide by meiosis, they have to make sure that each offspring gets one of each kind of chromosome. By matching up the pairs at the beginning, cells can make sure the pairs separate correctly. When the homologous chromosomes are stuck together, they can exchange pieces of their DNA in a process called *crossing-over* (see Figure 5-3A), resulting in new combinations of genes in offspring. The mitotic spindle attaches to the chromosomes just as it does in mitosis.

   ● **During metaphase I,** the mitotic spindle pulls homologous pairs to the middle region of the spindle. This occurs the same way as it does during mitosis, except that the chromosomes remain paired.

   ● **During anaphase I,** the mitotic spindle pulls one of each pair to opposite sides of the cell. The sister chromatids of each chromosome remain together.

   ● **During telophase I,** cytoskeletal proteins constrict and pull the nuclear membrane to form two nuclei, one around each set of chromosomes. At this point, each cell is haploid because it only has one of each kind of chromosome (even though those chromosomes still have sister chromatids).

>> **During meiosis II,** sister chromatids separate (see Figure 5-3B). Meiosis II proceeds through prophase II, metaphase II, anaphase II, and telophase II. The main events in each of these phases are the same as they are in mitosis. One important difference is that cells going through meiosis II are haploid, so they have half as many chromosomes as the parent cell.

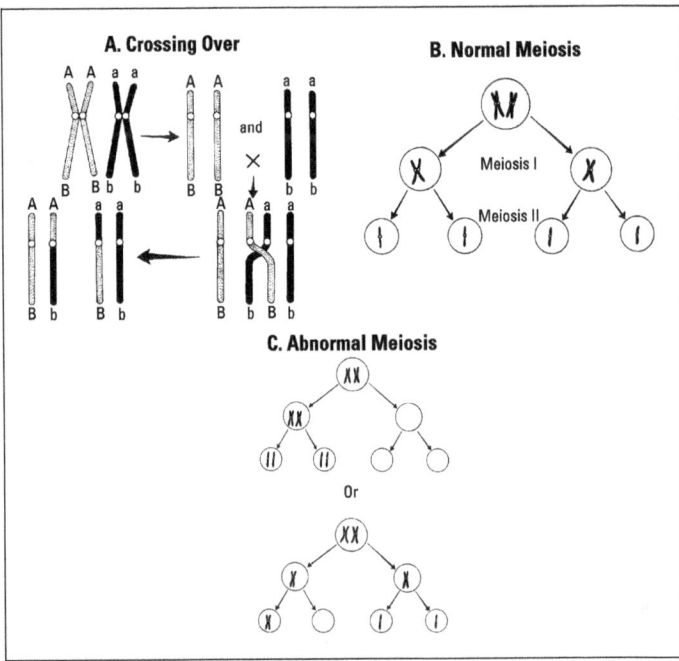

FIGURE 5-3: An overview of meiosis.

REMEMBER

One of the biggest differences between meiosis and mitosis is that homologous chromosomes pair up during meiosis I, but not at all during mitosis. The homologous chromosomes pair up during prophase I, line up in pairs during metaphase I, and separate from each other during anaphase I.

Identifying Sources of Variation

As they grow and reproduce, fungi can change their DNA by several different methods. The rate of change increases when individuals experience environmental stress, suggesting that these changes might even help individuals survive in the short term. In the long term, genetic changes that get passed on to the next generation can lead to the evolution of new strains and species of fungi.

Nonsexual variation

All cells acquire changes in their DNA over time. The enzymes that copy DNA aren't perfect, so whenever cells reproduce, a few mistakes are made.

The consequences of these changes range from absolutely no effect to cell death to cancer in multicellular organisms. In diploid cells, a serious mutation that disrupts a gene may not have an effect because diploid cells have two copies of every gene. So even if a mutation makes one copy nonfunctional, the other copy can produce a functional protein.

Most fungal nuclei are haploid, however, so mutations are more likely to immediately affect cell function. In some cases, this may enable an individual to adapt quickly to environmental conditions. A successful individual may grow faster than the competition, leading to rapid changes in the characteristics of a population.

Although most fungal nuclei are haploid, many filamentous fungi have more than one nucleus in the cell. These fungi may be able to change rapidly, while also being protected by having extra copies of each gene in a multinucleate cell.

Heterokaryons have two or more genetically different nuclei in the same cytoplasm.

REMEMBER

Genetic differences in the nuclei of heterokaryons can occur if one nucleus acquires a spontaneous mutation or if two different strains fuse and exchange nuclei. As the cells divide to produce the mycelium, different parts of the mycelium may contain different ratios of the various nuclei. The traits of these areas can differ as a result of the interaction of the genes in the local nuclei, giving the fungus more options for finding the resources it needs to grow.

Some heterokaryons reconfigure their genes even further by participating in something that's almost, but not quite, like sexual reproduction.

The *parasexual cycle* is an asexual process that transfers new combinations of genetic information to cells without meiosis or the formation of sexual structures.

The parasexual cycle begins when two of the haploid nuclei inside a heterokaryon fuse, creating a diploid nucleus. The diploid nuclei multiply by mitosis, during which crossing-over may occur (see Figure 5-3A). Over time, the nuclei may lose chromosomes as a result of nondisjunction (see Figure 5-3C), eventually becoming haploid again. Each nucleus that undergoes nondisjunction could lose different groups of chromosomes, resulting in genetically different nuclei after division.

Nondisjunction occurs when chromosomes fail to separate properly during cell division. It leads to *aneuploid* cells that have an incorrect number of chromosomes.

Nondisjunction can occur if cells fail the M checkpoint, causing cells to proceed through division before all the mitotic spindle captures all the chromosomes.

Some fungi that combine nuclei via the parasexual cycle have never been observed to do sexual reproduction, so this cycle may be an important way for them to generate genetic variation. The parasexual cycle can result in heterokaryons with very diverse populations of nuclei.

Sexual variation

Sexual reproduction produces genetically diverse offspring. For organisms that live in changing environments, it increases the likelihood that some offspring will survive. In some fungi, environmental stresses like limited food can increase the likelihood of sexual reproduction.

Sexual reproduction increases genetic variability by three mechanisms:

>> **Crossing-over:** During prophase I of meiosis, homologous pairs of chromosomes stick together. While they are in close contact, they may swap pieces of DNA between their sister chromatids as shown in Figure 5-3A, leading to combinations of gene variations that didn't exist in either of the parents.

>> **Independent assortment:** During meiosis I, each pair of homologous chromosomes sorts independently with respect to the other pairs. In other words, when one pair of homologous chromosomes separates and goes to opposite poles of the cell, it doesn't affect how any other pair separates. This means that every time a diploid parent produces haploid cells, they randomly distribute the copies of each chromosome.

>> **Random mating:** If the same two diploid parents produced numerous haploid cells that then fused together to form a new diploid organism, each new organism could come from a combination of genetically different haploid cells.

2

Diving Deeper Into Fungal Groups

Discover the importance of fungi to genetics and how examining the genetics of fungi has helped us learn more about our own DNA.

Take a look at the origins of fungi, including fossil fungi and fungi's ancestral groups.

Explore the zoosporic fungi and discover how they provide clues to the early evolution of Kingdom Fungi.

Examine the zygomycetous fungi, which include the familiar black bread mold and the fungi that grow on animal dung and dead plant material.

Check out the characteristics of the Ascomycota fungi, which include black and green molds, baker's yeast, powdery mildews, morels, cup fungi, and truffles.

Review the diversity of the Basidiomycota fungi — the puffballs, stinkhorns, bracket fungi, rusts, smuts, jelly fungi, and more.

Chapter **6**

Mapping the Genetics of Fungi

ungi make great organisms for genetic studies for a number of reasons. They are small and can be grown easily in the laboratory. They grow rapidly and have short life cycles that produce new generations quickly. Most fungi are *haploid,* meaning they only have one copy of each gene in their nuclei. This allows scientists to see the impact of genetic changes because there isn't a second copy of the gene to hide any mutational effects. They have relatively small amounts of DNA per cell, which makes it easier for scientists to decipher their genetic code. This chapter introduces you to how examining the genetics of fungi has helped us learn more about our own DNA.

Exploring the Genomes of Fungi

DNA, short for *deoxyribonucleic acid* (see Figure 6-1), is the hered-itary molecule of all cells on Earth, including those of fungi. DNA contains the instructions for building the molecules that deter-mine the traits of living things. Your particular blueprints, called genes, are a large part of why you are the way you are, from whether you can digest lactose to whether your hair is naturally straight or curly.

A *gene* is a blueprint that tells a cell how to build a worker molecule. These workers are usually proteins but can also be RNA.

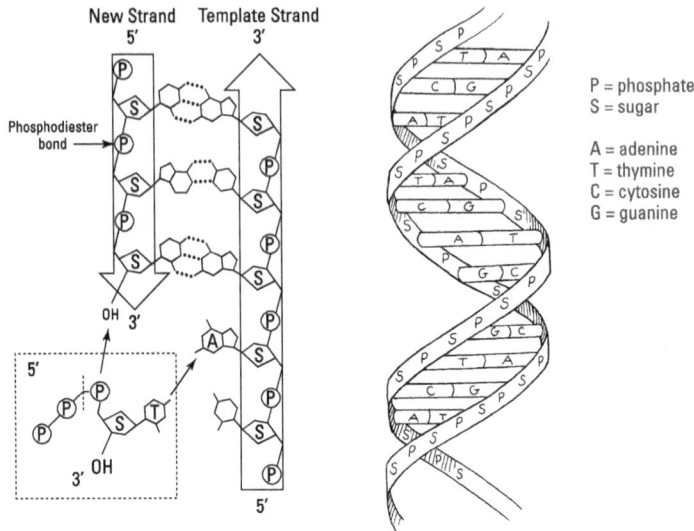

FIGURE 6-1: Deoxyribonucleic acid (DNA).

The genetic code in DNA is spelled out in chemicals called *nucleotides.* (There's a box drawn around one nucleotide that's about to be added to the DNA strand in Figure 6-1.) Four nucleotides make up our DNA, each with a different *nitrogenous base.* Scientists refer to the four nucleotides found in DNA by the first letter of the name of the nitrogenous base found in the nucleotide (you can see these letters in the DNA on the right in Figure 6-1):

>> A for adenine

>> T for thymine

>> C for cytosine

>> G for guanine

Cells join nucleotides to form long chains like those shown in Figure 6-1. Each DNA molecule consists of two chains held together by weak electrical attractions called *hydrogen bonds* that form between the nitrogenous bases. (On the left side of Figure 6-1, you can see the hydrogen bonds shown as dashed lines between two strands of DNA.)

The hydrogen bonds only form between specific pairs of bases (scientists refer to this as the *base-pairing rules*):

>> A always pairs with T
>> C always pairs with G

Scientists often refer to a *twisted ladder model* to explain the structure of DNA (shown on the right in Figure 6-1). If you think of a DNA molecule like a ladder, the bases from the two strands attach to each other to form the "rungs" of the ladder (shown as A-T and G-C). The "handrails" of the ladder consist of alternating sugar molecules and phosphate groups (shown as S and P) from the nucleotides that were joined together to make the long chains. Scientists call this the *sugar-phosphate backbone* of DNA.

Just as you have genes that determine your traits, fungi have genes that determine theirs. Spores from a particular type of mushroom contain the DNA of that species, so they grow and eventually produce the same type of mushroom as the original individual.

A *genome* is all the genetic information of an organism.

People and fungi are both eukaryotes (see Chapter 4 for more on the eukaryotic cell), so the bulk of the genome is contained within the nucleus of the cell. Mitochondria, located in the cytoplasm of the cell, also contain DNA that contributes to cellular function. Scientists often refer to this separately as the *mitochondrial genome*.

The types of DNA in a genome can be separated into categories based on function:

>> **Coding DNA** is the DNA that's found within genes. In other words, the DNA that cells actually use as blueprints to build molecules.

>> **Non-coding DNA** is DNA that isn't typically used as a blueprint. These are stretches of DNA that are often involved in how and when a cell uses a gene.

Scientists measure the size of genomes in *base pairs* (bp). Each base pair is one "rung" of the DNA ladder (A-T or G-C). DNA molecules can be very long, so it's sometimes easier to talk about kilobases (kb=1,000 bp), megabases (Mb=1,000,000 bp), and gigabases (Gb=1,000,000,000 bp).

We use the same prefixes to talk about computer memory. A byte is a small unit of memory, so we talk about kilobytes (KB=1,000 bytes), megabytes (MB=1,000,000 bytes), and gigabytes (GB=1,000,000,000 bytes).

Genome size is also sometimes reported by the actual weight of the DNA in picograms, or pg.

The genomes of fungi are smaller than those of most other eukaryotes. For example:

>> **The human genome is considered to be very average for animals, with a size of about 3.2 GB (3,200,000,000 bp).** Animal genomes vary widely, with some of the smallest found in roundworms (19.6 Mb in the parasitic species *Pratylenchus coffeae*) to some of the largest in species of lungfish (133 Gb in *Protopterus aethiopicus*).

>> **The *Arabidopsis thaliana* genome is about 135 Mb (1,350,000bp), putting it close to the middle of the pack of widely ranging plant genomes.** *A. thaliana* is a small mustard plant that's used for research on plant development and genetics. Across all land plants, genomes range from 61 Mb in the corkscrew plant (*Genlisea tuberosa*) to 160 Gb in the fork fern (*Tmesipteris oblanceolata*).

>> **Two fungi used extensively in genetic research, the bread mold, *Neurospora crassa*, and baker's yeast, *Saccharomyces cerevisiae*, have genomes of about 43 Mb (43,000,000 base pairs) and 12 Mb (12,000,000 base pairs), respectively.** Fungal genomes range from the smallest of about two Mb in the parasitic microsporidian (*Encephalitozoon intestinalis*) to the largest of about 5.8 Gb in the orange-yellow cup fungus (*Neottiella rutilans*).

The small size of fungal genomes is an advantage for scientific research. One of the first steps in studying a genome is to use *DNA sequencing* to read the order of As, Ts, Cs, and Gs in each chromosome. Smaller genomes require fewer resources and time to obtain a complete sequence.

The first eukaryotic genome to be completely sequenced by scientists was that of baker's yeast, *Saccharomyces cerevisiae*.

Once a genome is sequenced, scientists can begin to explore the meaning of the genetic code. They look for markers that identify genes and try to identify the molecule made from the gene. They also examine the non-coding DNA to learn more about how DNA is regulated. Studying fungal genomes with these approaches not only helps scientists better understand fungi themselves, but it also provides comparisons that improve understanding of genome structure and function in other organisms. In particular, studies of the genomes of baker's yeast (*Saccharomyces cerevisiae*) and the fission yeast (*Schizosaccharomyces pombe*) led to breakthroughs in the scientific understanding of the cell cycle (see Chapter 5) and the regulation of gene activity.

Examining Fungal Chromosomes

The DNA of most fungi is organized into linear chromosomes.

REMEMBER

Most fungi spend the majority of their life as haploid cells that only have one copy of each chromosome per nucleus (for a refresher on haploid versus diploid, head back to Chapter 5).

The number of chromosomes in haploid fungal cells ranges between 3 and 40, with most fungi having between 6 and 16. In the bread mold *Neurospora crassa*, n=7, whereas in baker's yeast (*Saccharomyces cerevisiae*), n=16.

Each chromosome contains several types of information:

>> **Genes** contain the instructions for building worker molecules:

- **Proteins perform many functions for cells.** To give just a few examples, proteins called *enzymes* speed up chemical reactions, *cytoskeletal proteins* support cells and help things move (see Chapter 4), *DNA-binding proteins* help regulate DNA, and *receptor proteins* receive signals. Without proteins, cells simply couldn't function and maintain life.

- **RNA molecules play important roles in information processing in cells.** Their structure is very similar to that of DNA (RNA stands for ribonucleic acid). Like DNA, they use the chemical pattern of nitrogenous bases as information. *Messenger RNA* (mRNA) molecules carry the blueprints for protein construction from the nucleus of the cell to the ribosomes in the cytoplasm (see Chapter 4 for more on ribosomes). *Ribosomal RNA* (rRNA) and *transfer RNA* (tRNA) both play roles in protein construction. *Small RNAs* (sRNAs) defend fungal genomes from viral damage and help regulate genes.

>> **Regulatory DNA** are sequences of DNA that interact with cellular elements to control genome structure and regulate gene activity. They are non-coding, meaning they aren't used as blueprints to build something. They perform their function as DNA code.

- **Each chromosome has a *centromere*,** which is a section of DNA recognized by cytoskeletal proteins that move chromosomes around during cell division.

- **The tips of each chromosome end in *telomeres*,** repetitive sequences of DNA that protect chromosomes from being broken down by the cell.

- **Each chromosome has multiple *origins of replication* (ORI),** which mark starting locations for the enzymes that copy DNA.

Variation in fungal genomes

Scientists studying fungal genomes have found differences in closely related species and even within the same species. These differences include:

>> **Chromosomal rearrangements:** Pieces of chromosomes may be missing, duplicated, or moved from one location to another, sometimes within a single chromosome (intrachromosomal) or between two chromosomes (interchromosomal).

>> **Different numbers of chromosomes:** Some species have extra chromosomes that contain collections of genes that improve their ability to be pathogenic.

Another thing that can induce change in genomes is the presence of mobile genetic elements.

REMEMBER

Mobile genetic elements are pieces of genetic material that can move within the genome of an organism and sometimes even move from one organism to another. Scientists also call these pieces of DNA *selfish genetic elements* because their function may be only to reproduce themselves.

REMEMBER

Fungal genomes are variable, even between members of the same species.

Scientists are studying the mechanisms behind these genome changes and the origins of extra chromosomes. Some scientists think the extra chromosomes form when existing chromosomes break. These broken pieces of DNA could then be passed to some cells during cell division (for more on cell division, flip back to Chapter 5).

One gene, one enzyme

Our modern understanding of genes owes much to the red bread mold (*Neurospora crassa*) and two clever scientists named George Beadle and Edward Tatum. After an English doctor named Sir Archibald Gerrard showed that certain metabolic defects could be inherited within families, Beadle and Tatum used *Neurospora crassa* to figure out how the inheritance of metabolism worked:

1. First, the scientists created mutants of *N. crassa* that required nutritional supplements to grow.

2. Next, the scientists sorted the mutants into groups based on which supplement each strain needed.

3. Finally, they tested mutants within each group to see whether their ability to grow was affected by the same or different genetic changes.

For example, Beadle and Tatum compared mutants that required the amino acid arginine to grow. They hypothesized that each mutant may have a different defect that stopped them from being able to make arginine for themselves. When they compared three mutants in this group, they found that two of them could also be rescued by the addition of the amino acids ornithine or citrulline (as shown in Table 6-1).

TABLE 6-1 **Beadle and Tatum Experiment**

Mold Strain	Minimal Medium	Minimal Medium plus Ornithine	Minimal Medium plus Citrulline	Minimal Medium plus Arginine
Wild Type	Grew	Grew	Grew	Grew
Mutant 1	Did not grow	Grew	Grew	Grew
Mutant 2	Did not grow	Did not grow	Grew	Grew
Mutant 3	Did not grow	Did not grow	Did not grow	Grew

By analyzing this growth pattern, Beadle and Tatum figured out that each mutant had a mutation in a different gene that coded for enzymes needed in a chemical pathway that produces arginine, as shown in Figure 6-2.

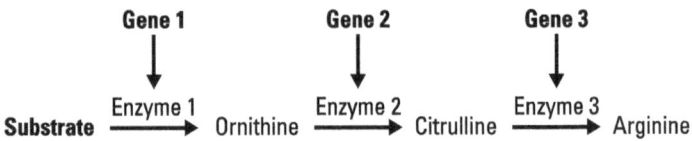

FIGURE 6-2: The one gene, one enzyme hypothesis.

Beadle and Tatum concluded that what is inherited from one generation to the next in mold and in people is the instructions to make enzymes. They proposed that each gene controls a specific enzyme. This groundbreaking idea became known as the "one gene, one enzyme" hypothesis. It connected almost one hundred years of genetic research, making sense of Dr. Garrod's work with metabolic diseases in families and the "heritable factors" first proposed by the Austrian monk Gregor Mendel in 1866.

The hypothesis of Beadle and Tatum has stood the test of time, with a few adjustments as geneticists made additional discoveries. When scientists learned that genes could contain instructions for other types of proteins besides enzymes, the idea was modified to "one gene, one protein." Today, we also know that genes contain the codes for functional RNA molecules that affect traits without ever becoming proteins, so we've had to update the idea again. It doesn't sound quite as catchy to say "one gene, one functional molecule," but it is more accurate.

Mapping Chromosomes

Our modern understanding of genetic diseases and how they are inherited from one generation to the next is supported by information gathered in chromosome maps.

REMEMBER

A *chromosome map* shows the position and relative distance between genes on a chromosome. Chromosome maps help scientists understand how a genome is organized.

Scientists use two main techniques to create chromosome maps:

>> *Genetic mapping,* or linkage mapping, determines the relative distance between two genes or other markers on a chromosome based on how often they are inherited together. It's based on the probability that markers that are closer together are more likely to be inherited together than markers that are farther apart. Scientists do genetic crosses between organisms to observe how often traits travel together versus independently.

>> *Physical mapping* involves techniques that visualize or manipulate physical chromosomes to determine the locations of genes and other chromosome markers.

Genetic mapping in fungi

Genetic mapping in fungi is helped by the fact that most fungi are haploid, which makes it easier to track the inheritance of particular alleles.

REMEMBER

An *allele* is a version of a gene.

For example, in the Beadle and Tatum experiments described earlier in this chapter, *Neurospora crassa* had genes for enzymes necessary to make the amino acid arginine. For each gene, the individual molds either had the wild type allele to make a functional enzyme or the mutant allele for a nonfunctional enzyme.

Ascomycete fungi (Chapter 10), such as yeast and the red bread mold, make genetic mapping easier because they package up the products of meiosis (Chapter 5) in a sac called an *ascus,* like the ones shown in Figure 6-3. Scientists collect the asci and dissect each one to obtain the spores produced from a single meiotic event. In yeast, the four products of meiosis form a *tetrad* within the ascus. In filamentous fungi like the red bread mold, the original spores of the tetrad each divide again by mitosis (Chapter 5), producing an *octad* of spores.

Genetic mapping began with the research of Thomas Hunt Morgan on fruit flies in the early 1900s. Morgan noticed that some combinations of traits seemed to be inherited together. For example, when he crossed mutant flies with miniature wings and white eyes with normal flies, the offspring almost always had *parental combinations* of these traits: that is, most of the offspring had miniature wings and white eyes or normal wings and normal eyes, with very few flies having *recombinant* appearances (miniature wings, normal eyes or normal wings, white eyes). These results were startling because they didn't match what geneticists knew at the time based on Gregor Mendel's research on peas. In peas, traits such as color and shape seemed to be inherited independently of each other.

The *Law of Independent Assortment* says that genes don't affect each other with regard to how alleles sort during meiosis. When two traits are inherited independently of each other, parental and recombinant appearances show up in predictable ratios in offspring.

In Morgan's experiments with fruit flies, the appearances of the offspring didn't match the ratios predicted by the Law of Independent Assortment, which meant that the genes were somehow influencing each other during meiosis. This led Morgan to the idea that the genes for these two traits must be close enough on the same chromosome that they traveled together during meiosis and were rarely separated by crossing-over. Morgan and his students decided that they could figure out how close different genes were on a chromosome by looking at how often crossing-over occurred. In other words, they could use genetic analysis to create a map of a chromosome.

Linked genes are genes that are close together on the same chromosome. Their proximity makes it likely that they will be inherited together.

A. Parental yeast strains

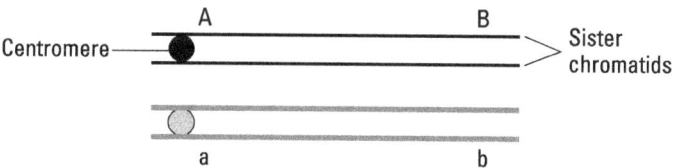

B. Possible outcomes and conclusions

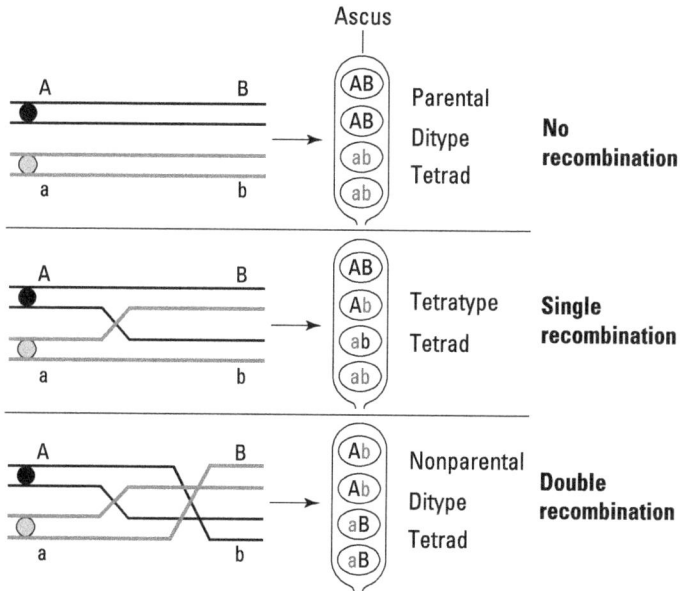

FIGURE 6-3: Tetrad analysis in yeast.

Beginning in the 1950s, a scientist named R.K. Mortimer began using genetic analysis to create maps of the chromosomes of yeast. Dr. Mortimer worked with other scientists over a period of 40 years to map all 16 yeast chromosomes. This information was combined with the physical map that was made after the yeast genome was sequenced (more on physical maps later in this chapter). The genetic maps are based on known traits, and they can be matched with areas of code from the physical map, ultimately helping to figure out the meaning of the genomic sequence.

REMEMBER

Scientists use the percentage of recombination events between two genes as a measure of the distance between them. The units given to these distances is the centiMorgan (cM) in honor of the famous geneticist Thomas Hunt Morgan, who first showed that genes are physically located on chromosomes. 1 cM is defined as 1% recombination.

ESCARGOT, ANYONE?

Brewer's yeast, *Saccharomyces cerevisiae*, has been an incredibly valuable organism for genetic research. It's small, inexpensive, and easy to grow in the lab. It's a eukaryote, so it has had lots to teach us about the structure and function of our own genes. In fact, scientists estimate that 30 percent of genes known to cause diseases in humans have analogous genes in yeast. Scientists created useful genetic maps of yeast chromosomes by the 1970s, and in 1996, yeast became the first eukaryote to have its genome completely sequenced.

Much of this amazing work on yeast may never have gotten done if it weren't for a clever biologist armed with some snail guts. In the 1950s, Robert K. Mortimer began using genetic analysis of yeast tetrads to create maps of the yeast chromosomes, but he ran into a problem. While many ascomycete fungi release their spores easily, the asci of yeast are tough stuff. Dr. Mortimer was finding it very difficult to break open the asci using the standard dissecting needles of the time. He and a colleague, John R. Johnson, looked around for something to help penetrate the ascus wall. Enter the common garden snail, which could quite literally digest the cell wall of yeast. Mortimer and Johnson extracted digestive juice from the snails, added it to water, and squirted it on the asci. The asci walls broke down, but the tetrads of spores stayed together. Genetic research in yeast became much easier for everyone, except, of course, the snails.

Physical mapping in fungi

REMEMBER

Physical mapping creates chromosome maps that show the exact position of DNA sequences on a chromosome. The distance between sequences is measured in base pairs.

Scientists use a number of different methods to create physical maps:

>> **One of the oldest methods is to stain chromosomes with dye and create *an ideogram*, which is a map of the physical features of the chromosome.** The pattern of stripes, called *bands,* on each chromosome is distinct and unique to each chromosome. Physical features of each chromosome, including the position of the bands, centromere, and telomeres, are used to create a numbering system for each chromosome. By comparing the appearance of normal chromosomes with those from someone with a disease, it's sometimes possible to locate a visible physical defect on one of the chromosomes, which helps identify where genes connected with that disease might be located.

>> **Another older technique is to use enzymes called *restriction enzymes* that cut DNA at specific sequences.** By treating DNA with a particular restriction enzyme and then looking at the resulting lengths of DNA fragments (called *restriction fragment length polymorphisms* or RFLPs), scientists could figure out where on a chromosome the enzyme made its cut. These cut sites, called *restriction sites,* could be marked on chromosome maps.

>> *Fluorescence in situ hybridization* **uses small pieces of DNA called *probes* to locate the position of particular sequences on a chromosome.** Scientists build probes with known DNA sequences and attach them to a molecule that glows when activated with a certain type of light. The scientists stick chromosomes to a glass microscope slide, then treat them with chemicals so the DNA in the chromosomes is looking for new DNA to partner with. They add the probes to the slide, and the probes stick when they find their matching sequence. For example, a probe with the sequence "AACC" would stick to a piece of a chromosome with the sequence "TTGG" (actual probes are much longer). The scientist shines the correct type of light on the slide and the probes begin to glow, showing exactly which chromosome has the matching sequence and where it is located.

>> **Sequence-tagged site mapping (STS) looks for unique DNA sequences in samples of DNA.** Scientists first have to choose an STS they want to look for; for example, they might know part of the DNA code for a particular protein and want to screen different DNA samples to see if the code is present.

One way to do this is to use a technique called *polymerase chain reaction*, which can rapidly make billions of copies of a piece of DNA as long as you provide the correct starter (called a *primer*). So, a scientist could build a primer that would allow their STS to be copied by PCR, then test different DNA samples to see which ones worked — in other words, which ones had the STS recognized by the primer.

Scientists have read the genomes of over 20,000 species of fungi, and many more are currently being worked on. For example, the Joint Genome Institute (JGI) of the United States' *1000 Fungal Genomes Project* aims to sequence the genomes of fungal groups that aren't currently represented in existing genome databases. These databases represent a powerful reference library that scientists can use to figure out the identity and function of newly discovered fungal species.

Chapter 7

Digging Into the Origins of Fungi

M ost scientists agree that the ancestors of today's mush-rooms and molds lived in the water and that they made their way onto land at about the same time as plants, although the exact timing of these events is still being debated. Because the fossil record of fungi is incomplete, scientists use comparisons of DNA sequences to try and sketch out the early evolution of fungi and determine how ancient fungi interacted with other life forms. Understanding the role of fungi during the early days of Earth can provide insight into their importance today. In this chapter, you look at the methods scientists are using to try to answer questions about the early evolution of fungi.

Writing the Fungal Origin Story

Many mysteries surround the origin of fungi. Phylogenetic anal-yses suggest that fungi may have emerged as a distinct group as early as a billion years ago, but clear fungal fossils don't appear until 400 million years ago. To try and fill in some of the gaps, scientists look at how modern fungi interact with other organ-isms to envision the role of ancestral fungi. They also look at evolutionary innovations in fungi and other organisms to see if they correspond. Using these approaches, scientists have come up with several possible scenarios for the origins of fungi.

Considering aquatic origins

Scientists generally agree that the ancestors of fungi lived in the water. One of the main pieces of evidence comes from phylogenetic analysis. The early branching groups of fungi (Chapter 8) have swimming stages, whereas the terrestrial fungi form a monophyletic group that lacks flagella or swimming stages. From this, scientists infer that fungi began as aquatic organisms and that the ability to make flagella was lost just once in the common ancestor to the terrestrial fungi.

Scientists are still debating whether fungi originated in marine or fresh water:

>> **Evidence for marine origin:** The sister group to the fungi is the Nucleariida, a group of amoebae that are common in marine environments. Also, marine fungal species do exist today.

>> **Evidence for freshwater origin:** Several pieces of evidence show a strong connection between fungi and green algae:

- The evolution of fungal enzymes to break down plant molecules parallels the evolution of plants.

- Green algae have receptors that recognize fungi and trigger an immune response, suggesting that green algae have been in close association with fungi for a long time.

Scientists believe that green algae evolved in freshwater, so this suggests that fungi were in freshwater habitats, too. Also, although marine fungal species exist, phylogenetic analysis suggests most of these evolved from terrestrial species.

Scientists put these pieces together to create a possible origin story for the ancestor of fungi as a parasite of green algae that lived in a freshwater environment. It fed by engulfing its food, swam with a flagellum, and had chitin cell walls during at least part of its life cycle. This ancestor sounds very much like modern chytrid fungi (discussed in Chapter 8).

Moving to land

Plants and fungi probably moved onto land at about the same time, although scientists are still investigating which group colonized land first. Plants are a great source of carbohydrates for fungi, so it may be that fungi followed their food source. On the other hand, modern fungi help plants obtain water and minerals from soil, so it may be that fungi made it possible for plants to move onto the land in the first place.

Scientists have proposed at least three hypotheses for how fungi moved from water to land:

>> Fungi moved to land as parasites of green algae, which helped protect them from drying out. First, they developed small projections called rhizoids (see Chapter 8), which eventually evolved into true hyphae. In this scenario, fungi and plants evolved together.

>> Fungi moved to land as free-living microbes in the soil as part of a microbial crust that included bacteria, algae, and protists. The fungi that are associated with algae became the modern terrestrial lineages (presented in Chapters 9, 10, and 11). The fungi that preyed on protists retained many ancestral traits and became the zoosporic fungi (Chapter 8).

>> Fungi moved to frozen terrestrial areas as parasites of green algae. The changeable and diverse frozen environment favored the evolution of hyphae and other traits that allowed fungi to move to dry land.

Once the ancestors of fungi moved to land, they evolved traits that helped them survive in their new environment:

>> **Hyphae** are very good at finding nutrients in uneven habitats. Hyphae branch to form mycelia that fill available space, increasing the chance that nutrients will be located and absorbed.

>> The development of **septa** within hyphae allowed for better regulation of the flow of nutrients through hyphae and the ability to minimize loss due to damage.

>> **Fruiting bodies** protect developing reproductive organs and improve spore dispersal.

>> **Fungal enzymes** coevolved with food sources, such as plants. The ancestors of fungi probably had the ability to break down cellulose, which is present in the walls of green algae and non-woody plants. After plants evolved the ability to make wood that contained lignin, fungi evolved enzymes to break down lignin. (See Chapter 2 for more on how fungi break down wood.)

>> Some fungi evolved a **mycorrhizal** relationship with plants, establishing a mutually beneficial relationship that provided food for fungi and water and minerals for plants. (See Chapter 3 for more on mycorrhizae.) Without this relationship, plants may not have been able to colonize the land at all.

Discovering Fossil Fungi

Fossils provide a window into the past as scientists try to piece together the history of life on Earth. Unfortunately, fungal structures don't preserve well, so the fossil record of fungi is sparse. In cases where fossils show structures that look fungal, such as threads that could be hyphae, it isn't always possible to distinguish whether these are truly fungi or whether they are structures from similar-looking organisms.

Some of the most convincing fungal fossils come from the Rhynie Chert, a deposit of sedimentary rock that formed in the early Devonian around 410 million years ago (see Figure 7-1) in the part of the world that is modern-day Scotland. Scientists found structures that look distinctively like arbuscles inside a root-like structure from the fossil plant *Aglaophyton*. Arbuscles are formed by arbuscular mycorrhizae (Chapter 3) in the Glomeromycota (Chapter 9). The Rhynie Chert also contained fossils of chytrids (Chapter 8) attacking other fungi and an apparent ascomycete (Chapter 10) attacking a plant.

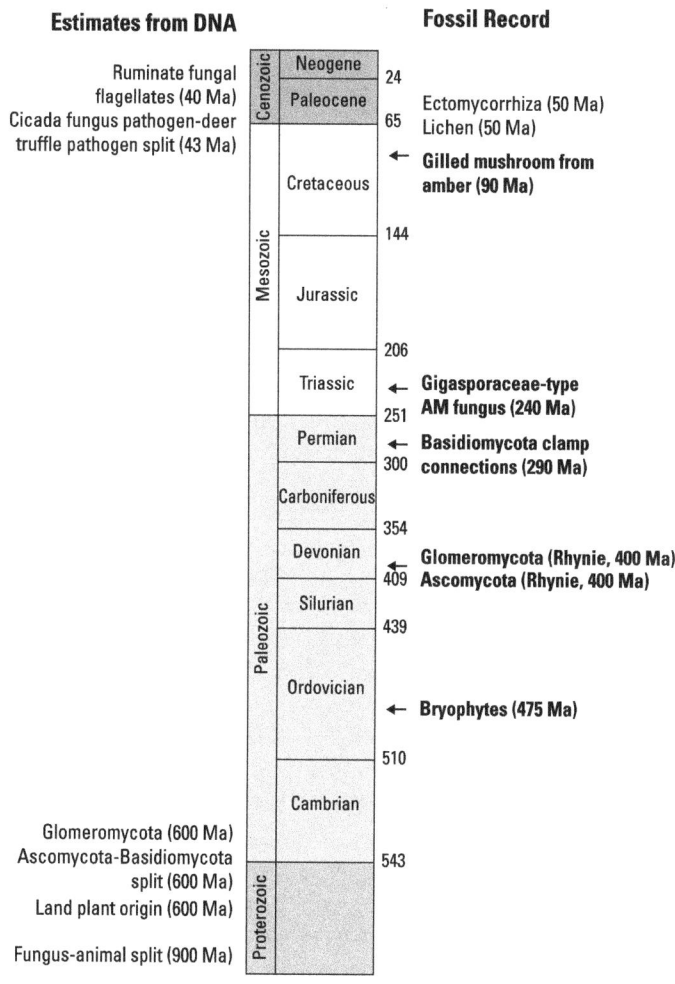

Estimates from DNA

Fossil Record

Ruminate fungal flagellates (40 Ma) Cicada fungus pathogen-deer truffle pathogen split (43 Ma)	Cenozoic — Neogene / Paleocene — 24 / 65	Ectomycorrhiza (50 Ma) Lichen (50 Ma)
	Cretaceous	← Gilled mushroom from amber (90 Ma)
	Mesozoic — 144 / Jurassic — 206	
	Triassic — 251	← Gigasporaceae-type AM fungus (240 Ma)
	Permian — 300	← Basidiomycota clamp connections (290 Ma)
	Carboniferous — 354	
	Devonian — 409	← Glomeromycota (Rhynie, 400 Ma) Ascomycota (Rhynie, 400 Ma)
	Silurian — 439	
	Ordovician — 510	← Bryophytes (475 Ma)
	Cambrian — 543	
Glomeromycota (600 Ma) Ascomycota-Basidiomycota split (600 Ma) Land plant origin (600 Ma) Fungus-animal split (900 Ma)	Proterozoic	

FIGURE 7-1: Fungal evolution based on the fossil record and DNA sequencing (Ma=millions of years ago).

Looking for Ancestral Groups

Scientists can supplement the information in the fossil record by building family trees of fungi based on the similarities and differences between species. In the past, scientists proposed relationships based on developmental, structural (morphology), and metabolic characteristics. For example, two fungi were considered closely related if they made the same type of repro-ductive structure. Likewise, parasitic species that lacked certain structures were grouped together or even excluded from the fungi altogether.

Scientists today have new tools that allow them to compare organ-isms down to the level of their molecules. Usually, this means going straight to the DNA code and comparing the sequences of particular genes (for more on genes and DNA code, head to Chapter 6). The ability to rapidly read DNA codes has dramati-cally changed the modern understanding of life on Earth for sev-eral reasons:

>> **It reveals invisible organisms.** Fungi spend most of their lives as barely visible hyphae growing through a substrate, making them almost impossible to identify by morphology alone. Now, biologists can extract DNA from substrates such as soil and look for genetic signatures to identify the organisms living in an environment. Not only has this revealed more detail on individual ecosystems, but it's vastly increased estimates of the number of microbes, including fungi, that are part of the biosphere.

>> **It allows the detection of organisms that can't be grown in the lab.** Some organisms require precise environmental conditions that can take years to figure out, and some, such as symbionts, may never grow in a lab.

>> **It allows identification independent of sexual reproduc-tion.** The morphology of sexual structures is a key feature for the identification of fungi, but scientists have been unable to observe these features in some species.

>> **It can confirm or reject identifications based on mor-phology.** Sometimes, two unrelated organisms have similar structures because of convergent evolution, making scientists think they're more closely related than they are.

TECHNICAL
STUFF

Unrelated organisms that live in similar environments may evolve similar characteristics. For example, plants that live in the desert typically conserve water by having thick stems and few leaves. As a result, *Agave* and *Aloe vera* look very similar to each other, even though they belong to two different plant families. When two dis-tantly related organisms independently evolve similar character-istics, scientists call it *convergent evolution*.

Making phylogenetic trees

Once scientists have collected information about the similarities between organisms, they use it to construct phylogenetic trees, like the one shown in Figure 7-2.

A *phylogenetic tree* shows the relationships and evolutionary history between groups of organisms.

Imagine drawing a family tree for your biological family, starting with yourself and moving back ten generations. The people closest to you in the tree, your siblings, children, or parents, will probably have more characteristics in common with you than the people who are ten generations in your past. In this example, you're drawing the tree based on knowing the relationships and then looking at the characteristics. Scientists do the reverse: They draw the tree based on the characteristics in order to figure out the relationships.

Scientists use computer programs to analyze their observations of structure, chemistry, and/or gene sequences of individual species or related groups called taxa.

A *taxon* (plural: *taxa)* is a group of related organisms in a biological classification system.

Computer software draws phylogenetic trees based on similarities between taxa, drawing the trees so that the branches of the most similar groups are closest together. Scientists can adjust the software so that it gives more importance to certain characteristics. For example, sexual structures are highly conserved, so they may be given more weight in an analysis.

A *conserved* gene, molecule, or structure is one that hasn't changed very much through evolutionary time. Conserved characters are typically essential to the survival of an organism. If changes occurred during evolution, they were more likely to lead to death and thus not passed on.

The best molecules for phylogenetic analysis have just the right amount of variability among the group you're trying to analyze. If there's too much variability, all the taxa will look different, and you won't be able to see relationships. If there's no variability, all the taxa will look the same. The perfect "Goldilocks" molecule differs depending on which group of organisms you're analyzing.

For fungi, the most commonly used DNA sequence for comparison is the one for ribosomal RNA (rRNA). rRNA makes up part of the structure of the ribosome, the organelle responsible for protein synthesis (more on ribosomes in Chapter 4). Because proteins are essential for the function and survival of all cells, the genes that contain the codes for rRNA are highly conserved.

A phylogenetic tree for fungi, based on rRNA analysis, is shown in Figure 7-2. One way to read the tree is to start at the left side, which represents the furthest point back in time. As you move from left to right:

>> The first branch point in the tree represents the common ancestor between fungi and animals.

>> The earliest branching groups within the fungi, like the Rozellidea and the Microsporidia, are thought to be the most ancient lineages within the fungi. Their exact position in the tree in relation to each other and the other older lineages, the Blastocladiomycota and the Chyridiomycota, is still being studied (these four groups are all presented in Chapter 8).

>> Next to evolve were the Zoopagomycota and the Mucoromycota (Chapter 9), followed by the Ascomycota (Chapter 10) and the Basidiomycota (Chapter 11).

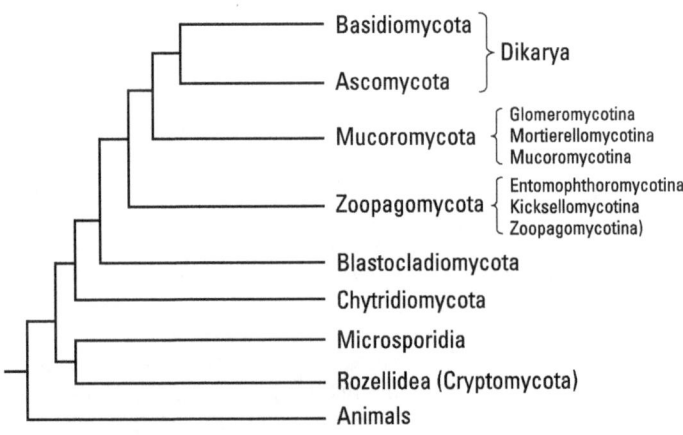

FIGURE 7-2: Fungal phylogeny.

The tree in Figure 7-2 shows the evolution of fungal lineages relative to each other, but it doesn't show when in the actual history of the Earth these events occurred. To place these events in geological time, scientists can calibrate a phylogenetic tree to known events such as those represented by fossils. By adding fossil data to a phylogenetic tree, with the assumption that fossils appear when a species is abundant, not when it first evolved, scientists can generate estimates of when evolutionary events occurred. The events and time estimates shown on the left in Figure 7-1, for example, are based on a calibrated phylogenetic tree.

Interpreting phylogenetic trees

To read a phylogenetic tree like the ones shown in Figures 7-2 and 7-3, look for the following information:

» The tips of the branches represent the species or other taxa that the scientist is comparing. Figure 7-3 shows a tree for three taxa: taxon A, taxon B, and taxon C.

» Branches meet at points called *nodes* that represent the common ancestor of the two taxa. Scientists call groups that branch out from the same common ancestor *sister groups*. In Figure 7-3, group B and group C are sister groups. In Figure 7-2, the Ascomycota and Basidiomycota are sister groups.

» An ancestor plus all its descendants form a *clade*. Figure 7-3 shows the clade for taxon B and C. In Figure 7-2, all of the fungal groups together form a clade.

   • If an ancestor and all of its descendants are in the same taxon, the taxon is *monophyletic*. Together, the Ascomycota and Basidiomycota in Figure 7-2 are part of a monophyletic group sometimes referred to as the Dikarya.

   • If a taxon contains an ancestor and some, but not all, of its descendants, the taxon is *paraphyletic*. A group of fungi discussed in Chapter 8, the Aphelidea, isn't represented on the tree in Figure 7-2. Although the placement of this group is somewhat uncertain, scientists currently think they share a common ancestor with the Microsporidia and Rozellidea. So, if the Aphelidea were placed into a different taxon than these two groups, the taxon would be paraphyletic.

>> Scientists call groups that branch from the base of the tree and are separate from the other groups *outgroups*. Scientists often deliberately include observations about a group that isn't very closely related to the group being studied in order to give a tree an outgroup. When the computer program has to include the outgroup, it helps give the tree scale by showing the group being studied in relationship to the larger picture of other kinds of life on earth. In Figure 7-2, animals are the outgroup to the fungi.

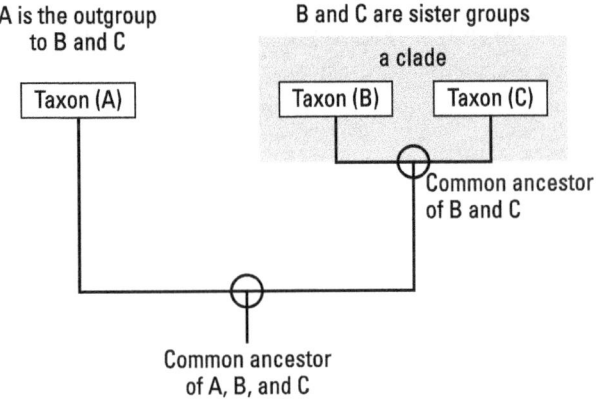

Maskot/Adobe Stock Photo

FIGURE 7-3: Reading a phylogenetic tree.

Creating a system

Scientists organize living things into taxa that reflect their relationships. Ideally, each taxon should be monophyletic.

Systematics is the study of the relationships among living things through time.

REMEMBER The more similar two organisms are, the more likely they are to be placed together in a category. The relative position of two organisms within the taxonomic hierarchy suggests the degree of their relationship. For example, you and a carrot are both in the domain Eukarya, so you definitely have some things in common, but you have more things in common with organisms that are also in the animal kingdom, like a dog. Table 7-1 gives you an idea of how the taxonomic hierarchy works by comparing the taxonomy of you, a dog, a carrot, and the fly agaric, *Amanita muscaria*.

TABLE 7-1 A Comparison of the Taxonomy of
Several Species

Group	Human	Dog	Carrot	Fly agaric
Domain	Eukarya	Eukarya	Eukarya	Eukarya
Kingdom	Animalia	Animalia	Plantae	Fungi
Phylum	Chordata	Chordata	Anthophyta	Basidiomycota
Class	Mammalia	Mammalia	Rosopsida or Eudicots	Agaricomycetes
Order	Primates	Carnivora	Apiales	Agaricales
Family	Hominidae	Canidae	Apiaceae	Amanitaceae
Genus	*Homo*	*Canis*	*Daucus*	*Amanita*
Species	*Homo sapiens*	*Canis familiaris*	*Daucus carota*	*Amanita muscaria*

REMEMBER

The taxonomic hierarchy for living things, from the largest, most inclusive group to the smallest, least inclusive group is Domain, Kingdom, Phylum, Class, Order, Family, Genus, Species.

Chapter **8**

Swimming with Zoosporic Fungi

W hen people think about fungi, they usually think of species that live on the land and visibly affect their lives, like mold or mushrooms. These fungi grow by hyphae that develop into mycelia (discussed in Chapter 4), and they reproduce by spores that blow in the wind. Before fungi moved onto land, however, they lived in the water. The common ancestor of fungi and animals was most likely a single-celled organism that swam with posterior flagella. Today, several groups of fungi still have characteristics that trace back to their early ancestors, including the production of asexual swimming spores called *zoospores*. In this chapter, you take a look at the swimming fungi and see what they have to tell us about fungal origins.

Reproducing with Zoospores

Organisms increase their chances of reproductive success by dispersing their offspring away from the parents. Essentially, offspring that move away from home have a better chance to thrive because they don't compete with parents or each other for the same resources. Species that live in water may take advantage of water currents, or they may have the ability to move easily by swimming.

When fungi and plants moved onto land, they had to develop strategies that used wind or animals to disperse their spores and seeds. One way that plants still show evidence of their aquatic origins is that they produce swimming sperm. Terrestrial fungi, on the other hand, completely lack any swimming stages. Because of this, scientists long debated whether any of the aquatic organisms that have fungal-like characteristics should be included in the fungal kingdom. The ability to read and compare the DNA of organisms has helped to resolve this issue by demonstrating that some fungal-like organisms with swimming stages are very closely related to terrestrial fungi, and others are definitely not. Scientists refer to the aquatic species included in the fungal kingdom as *zoosporic fungi*.

A *zoospore* is a motile spore that swims using a flagellum.

Although several fungal-like organisms produce zoospores, the zoospores of true fungi have specific characteristics. One key difference between the zoospores of true fungi and those of other organisms is that fungal zoospores have a single, posterior, smooth flagellum. Scientists refer to smooth flagella that lack hairs as *whiplash* flagella. (The zoospores of other fungal-like aquatic organisms often have two flagella, one of which has hairs and is called a *tinsel* flagellum.)

The exact relationships between the groups of zoosporic fungi are still being studied using DNA analyses, so some of the groups described in this chapter may change in the future. Part of what makes these groups interesting is that they have some characteristics that link back to life in the sea and some that clearly link to terrestrial fungi. Thus, these fungi provide clues to the early evolution of Kingdom Fungi. In this chapter, I use group names based on recent DNA analysis and point out places where names have changed or relationships are still uncertain.

Living Off Others with the Opisthosporidia

The fungi in this group all make a living as parasites of other eukaryotes, entering host cells and taking what they need to grow and reproduce. The structures and strategies that these fungi use to conquer other organisms provide a window into the early evolution of the parasitic lifestyle that's widespread among the fungi.

Obligate intracellular parasites are organisms that can only reproduce inside the cells of other organisms.

Some of the organisms in this group have unusual characteristics or even lack certain features found in other fungi. This is likely a result of their evolution within their host organisms. Because they have access to so many resources from their host, abilities or structures that would be necessary for a free-living species could become nonfunctional without causing the death of the parasite. Over time, these characteristics could be lost from a lineage. The organisms in this group can be divided into three lineages: Aphelidea, Rozellidea, and Microsporidia.

Aphelidea

The aphelids are obligate parasites of small algae called *phytoplankton* that float in the surface level of fresh and salt water. Phytoplankton form the basis for aquatic food webs, so parasitism by aphelids may have impacts on the structure of aquatic communities. Scientists that extracted and sequenced environmental DNA samples found that aphelids seem to prefer fresh water to salt water and were particularly abundant in *eutrophic* water that had high nutrient levels.

Eutrophication of water occurs in response to high levels of nutrients, especially nitrogen and phosphorus. The nutrients cause increased reproduction of algae, called *algal blooms*, which can cause the water to become green and murky. When the algae die, decomposers use oxygen to break down the algae. This can lead to a drop in oxygen levels in the water and cause the death of fish and other animals.

Aphelid reproduction probably increases along with that of their prey during algal blooms. As primary consumers of these algae, aphelids may contribute to the changes that occur during eutrophication.

Aphelids have unique behaviors usually seen in amoebae, suggesting they might be the oldest lineage of the opisthosporidians. In addition to typical fungal zoospores, they may produce *amoeboid spores* that retract their flagella and crawl instead of swim.

Amoeboid movement occurs when cells crawl along surfaces by pushing their cytoplasm in a particular direction, then pulling the rest of the cell along.

When an aphelid invades its host, it leaves its cell wall behind, sliding into the host as a *protoplast* with just its cell membrane as a boundary. The aphelid then digests the host cytoplasm by phagocytosis, a type of feeding commonly seen in amoebae. As the aphelid feeds, it grows into a large multinucleate cell called a *plasmodium*. When the plasmodium fills the entire host cell, the membrane and walls separate the nuclei into spores, and the cycle begins again.

Phagocytosis occurs when a cell eats large particles by engulfing them with its plasma membrane.

Rozellidea

Like the aphelids, the rozellids parasitize small organisms that live in aquatic environments. The favored hosts for rozellids, however, are other zoosporic fungi like the chytrids described later in this chapter. They also parasitize protists called water molds and a few species of algae. Like the aphelids, rozellids impact the populations of their host organisms and affect the flow of energy and nutrients through the aquatic environments in which they act as secondary consumers.

Also like the aphelids, rozellids leave their own cell wall behind when they enter the host cell, then eat the host cytoplasm via phagocytosis. The parasite grows to fill the space in the host cell. Some species in this group can trigger the host to form septa within their sporangia, creating more walled compartments for the parasite to inhabit. The parasite may become a plasmodium that divides into zoospores, or it may form thick-walled resting spores.

Microsporidia

Microsporidians, like the one shown in Figure 8-1, are obligate intracellular parasites of animals and some protists. In humans, they cause a disease called *microsporidiosis*. The signs and symptoms vary based on how the parasite enters the body and how far it spreads. The most common form of the disease is diarrhea caused by infection of the digestive tract by *Enterocytozoon bieneusi*, but other species can infect the eyes and muscles. The infection can even be fatal if it spreads through the body. People who are immunocompromised are at the highest risk for infection.

Microsporidians seem to have lost quite a few characteristics during their evolution. They have very small genomes and lack motile structures. Most species don't have mitochondria or the ability to obtain energy from food molecules, relying instead upon energy carriers from the host. They use a unique structure called a *polar filament* (shown in Figure 8-1) to penetrate into host cells. When a microsporidian cyst lands on a host cell, it germinates and rapidly ejects the polar filament (also called a polar tube) so that it penetrates into the host and delivers the parasite protoplast.

TECHNICAL STUFF

Because microsporidians lack mitochondria, scientists once thought they were descended from a very ancient lineage of eukaryotes that existed before eukaryotes acquired mitochondria. However, microsporidians do have an organelle called a *mitosome* that seems to be derived from mitochondria. Also, DNA analysis revealed that microsporidians do have mitochondria-related genes. Taken together, this evidence suggests that the ancestors of microsporidians once had mitochondria but that they were lost over time.

WARNING

Until recently, scientists referred to the Rozellidea and some related organisms as the Cryptomycota. When DNA analysis revealed how closely the Rozellidea and Microsporidia were related, however, they thought the organization of lineages needed to change to show that relationship. Scientists are still sampling and studying the organisms in these groups, so it's not entirely clear how these older fungal lineages will fit into the fungal family tree.

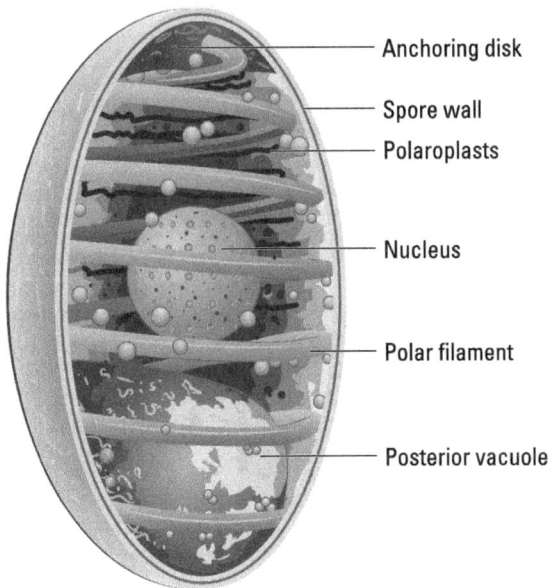

Anchoring disk

Spore wall

Polaroplasts

Nucleus

Polar filament

Posterior vacuole

FIGURE 8-1: Microsporidia.

Digesting the Tough Stuff with the Chytridiomycota

The phylum Chytridiomycota is the best studied of the zoosporic fungi, with over 1,000 species currently identified. Most species live as saprobes in aquatic habitats and soils, growing on decaying plant matter and pollen. A few species are parasites of plants, animals, and other fungi.

Currently, the two best-studied classes within the phylum Chytridiomycota are:

>> **Chytridiomycetes:** This is the largest class within the phylum, with about 1,000 described species. Most live in water, but some live in the soil. Many are saprobes, but a few are parasitic on plants and animals, including the species *Batrachochytrium dendrobatidis,* which is currently devastating populations of amphibians around the world.

>> **Monoblepharidomycetes:** These chytrids are saprobes that live in freshwater environments. Although the group is small,

with only about 35 currently described species, they are notable because some species produce true coenocytic hyphae similar to those of terrestrial fungi. They're also remarkable because, unlike all other fungi, they produce *anisogamous* gametes, with one type of gamete being larger than the other. As is typical with anisogamous gametes, scientists refer to the larger gamete as female and the smaller gamete as male.

Chytrids have several characteristics that are similar to those of terrestrial fungi, including:

>> Cell walls made primarily of chitin.

>> Filamentous growth in some species:

- Thin protoplasamic filaments called *rhizoids* that help anchor them to their substrate.

- Expansion of filamentous growth to form a *rhizomycelium*.

- Formation of true mycelia with *pseudosepta* that divide the hyphae.

TECHNICAL STUFF

The hyphae of rhizomycelia contain protoplasm but no nuclei, although nuclei may travel through them. The pseudosepta in species with true mycelia have a different chemical composition from the cell wall.

Scientists describe aspects of the chytrid life cycle based on how they grow and form zoospores (see Figure 8-2):

>> **Epibiotic versus endobiotic:** *Epi* means on, so epibiotic species grow on their substrate. *Endo* means inside, so endobiotic species grow inside the cells of their hosts.

>> **Holocarpic versus eucarpic:** *Holo* means entire, so holocarpic species convert their entire body into reproductive structures. For example, a holocarpic chytrid would convert its entire thallus plus rhizoids to make a zoosporangium. *Eu* means true, so eucarpic species only convert part of their body to make a sporangium. You can think of this like they have a true body that has both vegetative and reproductive structures. In these species, the rhizoids would not be incorporated into the zoosporangium.

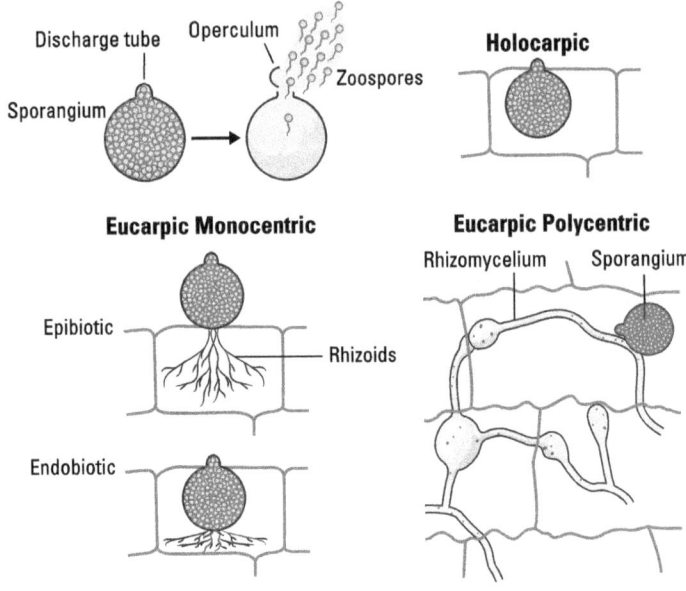

FIGURE 8-2: Types of chytrid thalli.

Discovering the many roles of chytrids in nature

Chytrids live in diverse aquatic and soil environments all around the world. They make significant contributions to the recycling of nutrients in their communities because they have the ability to digest molecules that other species can't. In addition, some chytrids live as parasites of plants and animals. In this section, you'll take a look at the various lifestyles of chytrids and how they impact the ecosystems in which they live.

Recycling nutrients

Like other fungi, chytrids are decomposers, producing enzymes that can digest the remains of other organisms. Chytrids can break down some of the toughest molecules nature has to offer, including cellulose from plant cell walls, chitin from insect exoskeletons and fungal spores, and sporopollenin from the outer wall of pollen grains.

When chytrids digest these tough molecules into smaller components, they make it possible for other organisms to access their available nutrients. Pollen, for example, is rich in fatty acids and minerals. Without digestion by chytrids and bacteria, however, it can settle to the bottom of lakes, remaining intact in sediments for thousands of years.

Chytrids can attach to pollen grains and complete their entire life cycle on their surfaces. Once chytrids penetrate the outer layer, other organisms, such as zooplankton like those shown in Figure 8-3, can access the pollen grain as a food source. Zooplankton also feed on the chytrid zoospores when they emerge from their sporangia. Chytrids make the nutrients from pollen available to the food web in two ways: by helping zooplankton digest the pollen directly, and by converting the nutrients in pollen into their own bodies, which are then eaten by zooplankton.

Chytrids also affect the availability of nutrients contained in algae, as shown in Figure 8-3. By starting the process of digestion on algae, they can make it possible for zooplankton to access the algae as food. Zooplankton can also feed on the chytrid zoospores when they emerge from sporangia on the algae.

REMEMBER

A *mycoloop* is a pathway of nutrient transfer in an aquatic ecosystem in which nutrients from inedible algae pass to zooplankton via the zoospores of chytrids.

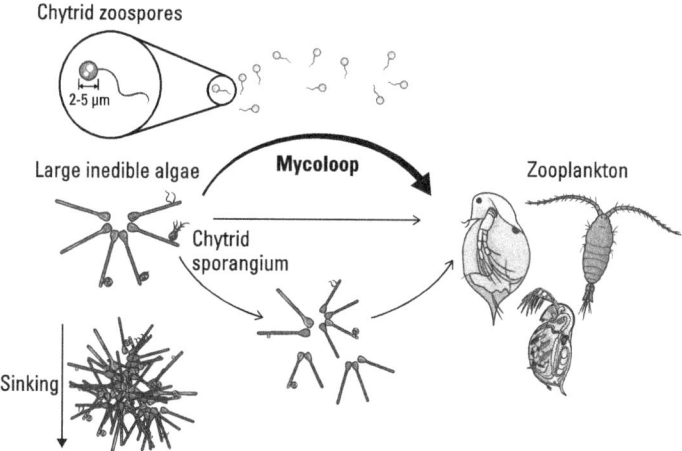

FIGURE 8-3: The mycoloop.

Attacking animals

Amphibians like frogs and salamanders around the world are dying in huge numbers due to infection by parasitic chytrids in the genus *Batrachochytrium*. The chytrids infect the skin of the amphibians, disrupting their ability to exchange gases with the atmosphere and affecting their electrolyte balance. The resulting disease, called *chytridiomycosis*, first causes skin discoloration and shedding, then ultimately leads to cardiac arrest. Scientists recently estimated that the fungi have driven at least 90 species to extinction and critically endangered another 124.

The deaths caught scientists' attention in the late 1990s, when amphibian populations in Central America, the Caribbean, and Australia started declining, and no one knew why. Scientists collected samples of dead frogs from sites where populations crashed and found *Batrachochytrium dendrobatidis* in the skin of the dead frogs. It took almost ten years for scientists to agree that this chytrid was responsible for the deaths, however, because of several factors:

>> Very few specimens were collected from the sites where amphibians had died, so some scientists felt like they didn't have enough evidence.

>> Some populations of frogs that had *B. dendrobatidis* on their skin had no signs or symptoms of disease.

>> The chytrid had lived for hundreds of years in some environments where deaths suddenly occurred, so scientists felt something new had to be involved.

>> The disease spread so quickly around the globe that some scientists couldn't believe it could be a waterborne pathogen.

After years of research didn't turn up any other likely possibilities, scientists agreed that *B. dendrobatidis* was the most likely culprit, and they started urging governments to take action to stop the spread. In nature, the chytrid spreads either by swimming zoospores or by skin-to-skin contact during mating. When scientists tested water that had been used to ship frogs around the world, they found the chytrid in the wastewater, suggesting that the global trade in frogs was contributing to the rapid spread around the world.

Although the reports of population collapse have been absolutely devastating, scientists have seen signs of recovery in some populations. Scientists are still studying the recovering populations, but it seems that some individuals are more resistant to the chytrid than others. They aren't infected as easily and, when they are infected, they recover more quickly. As these individuals reproduce, they may be passing resistance genes onto their offspring, leading to the evolution of more resistant populations.

Parasitizing plants

Many chytrids are pathogens or parasites of plants. For example, several species in the genus *Synchytrium* cause wart and gall diseases. *S. endobioticum* causes potato wart disease, also known as black scab. This disease is widespread in South America, especially in Peru, which is the place of origin of the potato plant. Scientists think this species also originated in Peru, then spread to Europe in the 1800s when new varieties of potato were introduced after the widespread potato blight.

Efforts to control potato wart disease have mostly focused on monitoring and containment, as the resistant spores have proven to be very difficult to kill, and they can remain alive in soil for decades. Farmers are switching to varieties of potato that show more resistance to the pathogen. Governments are testing soils for the presence of the chytrid and marking off safe zones where no pathogens are detected. Potatoes to be grown in these areas must be tested and proven to be free of the disease before they can be planted.

Taking a closer look at chytrid life cycles

The word chytrid comes from an Ancient Greek word that means "little pot." They got this name because, when scientists looked at them under the microscope, they thought chytrids looked like little pots full of spores. Chytrids have two growth stages, which are shown in Figure 8-4:

>> **The zoospore phase enables chytrids to disperse through their environment.** Like other fungi, chytrids produce zoospores with a single posterior whiplash flagellum. The zoospores detect and respond to both chemical

and light signals, moving toward food and light to find a good substrate for growth.

>> **During the growth phase, chytrids attach to their substrate and take the nutrients and energy they need to reproduce.** Unlike the opisthosporidians that use phagocytosis, chytrids practice absorptive nutrition similar to that of terrestrial fungi, secreting digestive enzymes into their substrate and then absorbing small molecules into their cells.

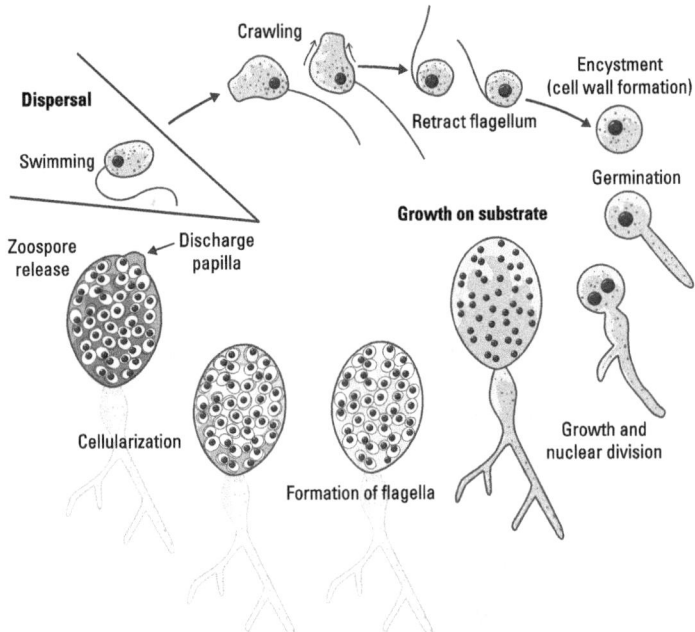

FIGURE 8-4: An overview of the chytrid life cycle.

Figure 8-4 shows the steps of the chytrid life cycle:

TECHNICAL
STUFF

1. Zoospores swim through the environment. Once they find a suitable place to grow, they settle onto a surface, retract their flagella, and form a cyst.

2. The cyst germinates, producing a germ tube that grows into the substrate. The rhizoids release digestive enzymes and absorb food molecules.

3. The chytrid grows, producing more nuclei by mitosis (for a review of mitosis, head back to Chapter 5) until it becomes a large, multinucleate cell.

4. A cell membrane and wall form around the nuclei, creating individual cells that form flagella to become zoospores.

5. The sporangium matures, and one or more small projections called *papillae* form in the sporangium wall.

6. The zoospores exit through the papillae, swimming away to find new substrate:

   a. *Operculate* chytrids make a small lid, called an *operculum* that will pop open on the discharge papillae to release the zoospores.

   b. *Inoperculate* papillae don't have a lid, instead releasing zoospores through pores or by dissolving the papillae.

Ruminating with Neocallimastigomycota

The Neocallimastigomycota is a small group of obligately anaerobic fungi. They live in the intestinal tracts of herbivores, where they help the animals digest their food. Herbivores eat plant material full of the complex carbohydrate cellulose, which animals can't digest by themselves. Microbes like these fungi digest the cellulose into smaller molecules that the herbivores can break down.

REMEMBER

Obligate anaerobes are killed by the presence of oxygen. In other words, they require the absence of oxygen (*an*=without, *aero*=air or oxygen).

Scientists are interested in the possibility of using these fungi to break down waste biomass, for example, to produce biofuels (for more on this process, check out Chapter 12). Not only do these fungi have the ability to make very large numbers of different enzymes to break down carbohydrates, but they also organize their enzymes into efficient packages called *cellulosomes*. The cellulosomes bring together enzymes that can work together to deconstruct the complex molecules found in plant tissue.

Attacking Crops with the Blastocladiomycota

The Blastocladiomycota is a small group of zoosporic fungi that live as saprobes in soil and freshwater and as parasites that attack plants, small animals, and other fungi. Some of the plant pathogens cause diseases in agricultural species, like brown rot in corn, crown wart of alfalfa, galls of sugar beets, and blight of citrus crops.

The blastoclads have some structural features that distinguish them from other zoosporic fungi. Some produce true mycelia with well-developed, branching hyphae. Scientists observed a visible cap-like structure near the nucleus in resting sporangia. Further investigation revealed that the *nuclear cap* was a dense cluster of ribosomes positioned near the nucleus.

Chapter **9**

Adapting to Life on Land with the Zygomycetous Fungi

The zygomycetous fungi, often called zygomycetes, is a group in the Kingdom Fungi that contains two phyla, the Zoopagomycota and the Mucoromycota. These fungi include the familiar black bread molds as well as various saprophytic species that grow on animal dung and dead plant material. Zygomycetes are used to produce foods such as tofu and tempeh. Some cause human disease, while others are parasitic on insects or fungi. A few form beneficial symbioses with plants and insects. In this chapter, you take a look at the structures that unite these fungi and find out more about their unique lifestyles.

Uniting the Zygospore-forming Fungi

Scientists organized the zygomycetous fungi into a group because almost all of them produce similar sexual spores. When compatible hyphae of opposite mating types meet, sexual reproduction may occur, producing thick-walled resting spores called *zygospores*. Zygospores may be pigmented and have outside textures such as bumps or warts.

Scientists using DNA to further explore the relationships within the zygospore-forming fungi are still making decisions about the best way to represent this part of the fungal family tree. The zygomycetous fungi, often called *zygomycetes,* are closely related, but the group itself is paraphyletic, so more changes will likely occur in the future. In this chapter, I'll use the most current groupings and point out places where the old organization has changed.

REMEMBER

The members of a *paraphyletic* group all have a common ancestor, but some of the descendants of that ancestor aren't included in the group.

Zygomycetes have characteristics that improve their chances for success on land. Unlike their aquatic relatives (discussed in Chapter 8), they don't produce flagella or require water for dispersal of their spores. Most grow by coenocytic hyphae that develop into mycelia, enabling them to penetrate extensively through substrates such as dead or living organisms.

REMEMBER

Coenocytic hyphae don't have very many septa, or cross-walls. They are like open straws filled with nuclei. (For more on septa, flip back to Chapter 4.)

WARNING

Most of the zygomycetous fungi used to be grouped together in a phylum called the Zygomycota. This phylum was defined by the production of zygospores, growth by coenocytic hyphae, frequent asexual reproduction by sporangia, and the lack of multicellular spore-producing structures called sporocarps (for example, mushrooms are sporocarps). DNA analysis demonstrated that this organization wasn't the best fit according to the evolutionary history of the fungi, so it's no longer valid.

Approximately 2,000 species of fungi are currently organized into this group. Scientists divide the group into at least two phyla, the Zoopagomycota and the Mucoromycota. The Zoopagomycota is relatively small, with only about 200 species. The ancestors of this group were the first fungi to colonize the land. Today, many of these fungi live in association with animals, either as parasites, pathogens, or commensal symbionts, while a few live as saprobes. The Mucoromycota is much larger, with over 1,000 species, and consists mostly of fungi that live in association with plants, either as saprobes or as symbionts.

WARNING

The Glomeromycota are a group of fungi that form arbuscular mycorrhizae with plants. (See Chapter 3 for details on these mycorrhizae.) Because scientists have never seen these fungi form zygospores, they were historically excluded from the zygomycetous fungi, and their relationship to the rest of the fungi was unclear. Based on DNA analysis, scientists now propose that these fungi are very closely related to other zygomycetous fungi. They are still debating whether these fungi will be organized into their own phylum or whether they will be included in the Mucoromycota. In this chapter, I include them in the Mucoromycota with the acknowledgment that this organization may change in the future.

Forming Mycelia with Zoopagomycota

The Zoopagomycota contains the groups of terrestrial fungi that appear to have evolved first. All of the fungi in this group form true mycelia. Most are either saprobes or parasites on small animals or other fungi. The phylum is divided into three subphyla that are grouped based on similarities in their DNA and cellular structures:

>> **The fungi in the Entomophthoromycotina all have a unique sterol molecule in their plasma membranes.**
   Scientists currently further divide them into three classes:

   - The Basidiobolomycetes include the genus *Basidiobolus*, which can cause opportunistic infections in humans. It's also unique among the fungi in this subphylum because it produces septate hyphae.

   - The Neozygitomycetes contains many parasites of mites and aphids.

   - The Entomophthoromycetes includes many parasites of insects (*entomon*=insect, *phthoro*=destroyer), roundworms and plants. *Entomophthora muscae* attacks flies, burrowing under their cuticle and digesting their bodies from the inside. The fungus can turn flies into zombies, taking over their bodies and causing the flies to climb to a high place before they die. The fungus pushes hyphae called conidiophores out through the cuticle, then shoots out asexual spores called *conidia* from the conidiophores. If the initial *primary conidia* don't land on a suitable host, they will try again, ejecting *secondary conidia* to a new location. The species *Entomophthora maimaiga* attacks certain species of moths and has had some success as a biocontrol agent to control gypsy moth infestations.

>> **Some members of the Zoopagomycotina form septa.** This subphylum consists mainly of species that live as parasites of amoebae, small animals, and fungi. *Zoophagus insidians* preys on small aquatic animals called rotifers (*zoo*=animal, *phagus*=eater). The fungus produces a mycelium that is covered with lollipop-shaped projections that trap rotifers. Bacteria and viruses attach to these projections. When the rotifers get stuck, the fungus penetrates into the animal, carrying bacteria and viruses in with its hyphae. The bacteria digest the rotifer from the inside while the fungus absorbs some of the food, until all that's left is a depleted shell.

>> **The fungi in the Kickxellomycotina have septate hyphae with holes called pores that allow cytoplasm to flow between cells.** The structure of the pores in this group is unique because they are surrounded by flattened disks called *lenticular plugs*. The group consists mostly of saprobes that live in soil and on dung. They form one-spored sporangiola. Most species are saprobes in soil or dung, but those in the genus *Martensella* are parasitic on other fungi.

Forming Arbuscular Mycorrhizae with the Glomeromycota

Almost all the fungi in the Glomeromycota are obligate symbionts of plants. They form arbuscular mycorrhizae, which are the most common type of mycorrhizae in the world. The fungal mycelium spreads through the soil around the roots of plants, absorbing water and minerals like nitrogen and phosphorus. The fungus also penetrates into the plant roots, forming structures inside plant cells that help it transfer water and minerals to the plant in exchange for sugars made by photosynthesis. For more details on arbuscular mycorrhizae, head back to Chapter 3.

Instead of forming arbuscular mycorrhizae with plants, the species *Geosiphon pyriforme* forms symbiotic associations with photosynthetic bacteria called cyanobacteria (or blue–green algae).

The hyphae of *Geosiphon* swell to form balloon-like bladders. The cyanobacterium *Nostoc* lives inside these swellings, making sugars via photosynthesis and capturing nitrogen from the air in the process of *nitrogen fixation*. *Geosiphon* gets food and nitrogen from *Nostoc*, providing protection from the sun in exchange. In a functional sense, this symbiosis is very similar to the lichens discussed in Chapter 3.

Scientists haven't observed sexual reproduction in the Glomeromycota, which makes it difficult to determine their relationship to other groups of fungi. Scientists agree that, based on their DNA, these fungi are very closely related to those in the phylum Mucoromycota. As more evidence becomes available, scientists may reorganize the fungi into these groups.

Getting Familiar with Mucoromycota

The Mucoromycota is the largest phylum of zygomycetes with over 1,000 species identified so far. Most of the fungi in this group are saprobes, but some are parasites, and a few form ectomycorrhizae (for more on ectomycorrhizae, head back to Chapter 3). Rarely, they can cause very invasive infections of humans and other animals.

Scientists divide this phylum into two subphyla: Mortierellomycotina and Mucoromycotina.

>> Fungi in the Mortierellomycotina don't produce a columella in their sporangia. Many of them grow as fluffy white colonies that produce a garlic-like odor. Species in the genus *Mortierella* are common in the soil where they grow as saprobes. Fungi in the genus *Aquamortierella* are aquatic.

>> Fungi in the Mucoromycotina do produce a columella in their sporangia. Scientists organize this subphylum into two orders:

- The **Endogonales** is a relatively small but important group of fungi, some of which form ectomycorrhizae with ancient lineages of plants such as liverworts, hornworts, and ferns. (For more on ectomycorrhizae, see Chapter 3.) They form small sporocarps below the soil, which can be a mixture of mycelia, zygospores, and sporangia. Some of these sporocarps are called "pea truffles" because of their resemblance to true truffles, but they are much smaller and not really edible. Species of *Endogone* are especially important in nutrient-poor soils such as sand dunes, where they help stabilize the sandy environment so that plants can establish themselves.

- The **Mucorales** is a large group of over 400 species. Most are saprotrophs, but some are parasites on plants. A few species are important to the food industry, including *Rhizopus oligosporus* and *Actinomucor elegans,* which are used to make tempeh and tofu, respectively. In the sections below, you will take a closer look at a few of the particularly well-known species in this order.

Spoiling food with Rhizopus

You may have already met the black bread mold *Rhizopus stolonifer* growing in your kitchen (see Figure 9-1). This species also grows as a parasite on plants, causing rot of ripe fruits such as strawberries and peaches.

Like most other zygomycetes, *Rhizopus stolonifers* grow by coenocytic hyphae, only producing septa when forming reproductive structures. It gets its name because it makes two specialized types of hyphae, which are shown in Figure 9-1:

>> *Rhizoids* are root-like hyphae that help reproductive structures attach to the substrate.

>> *Stolons* are lateral hyphae that connect two groups of rhizoids.

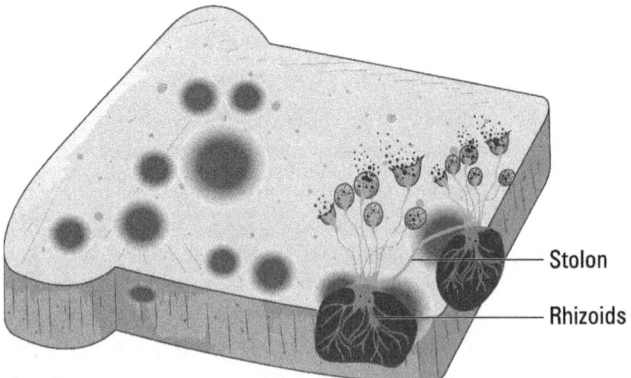

FIGURE 9-1: The black bread mold, *Rhizopus stolonifer.*

Zygomycetes are sometimes called "pin molds" because their asexual sporangia look like the straight sewing pins that have the ball-like tops. If you ever look at moldy bread and see a bunch of tiny black balls scattered across some fuzzy hyphae, you're probably looking at pin-style sporangia like those shown in Figure 9-1.

Detecting light with Phycomyces

Fungi in the genus *Phycomyces* are known for their ability to sense the environment. (One scientist even said they have all the senses we do except for hearing.) These fungi respond to light, detect air movements, and react to a variety of chemicals. Scientists have also observed an *avoidance response* in which the sporangiophores grow away from stationary objects. Although they hypothesize that the fungus figures out where objects are by releasing small amounts of a gas (think echolocation but with smell), no one has been able to figure out exactly how this works. When you combine all of these interesting behaviors with the fact that *Phycomyces* is relatively easy to grow in the lab and that it produces large sporangiophores, you can see why scientists might consider it to be a useful *model organism* in the lab.

A *model organism* is a non-human species that scientists study in order to figure out biological processes that are likely to have relevance to other organisms.

Phycomyces blakesleeanus helped scientists understand how organisms respond to light. One advantage of the fungus is that it uses light as a source of information and not for energy. Plants, on the other hand, use light for both. They have one set of pigments that absorbs light energy for photosynthesis, and other photoreceptors that react to light to get information about the time of day or the season of the year.

A *photoreceptor* is a molecule, usually a protein, that can absorb light. When photoreceptors absorb light, they pass a signal into a cell and trigger a response.

Many scientists over the years worked to understand how *Phycomyces* perceived and responded to light. In early experiments, scientists tested the growth of *Phycomyces* in response to different intensities and wavelengths of light. They determined that the mycelial growth reacts to light signals and that the sporangia of the fungus are positively phototactic, meaning they bend towards the light. They found that it was the blue portion of the spectrum that triggered these responses, which was exciting because blue light responses are important in other types of organisms as well.

REMEMBER

The sporangiophores of many zygomycetes show positive phototropism, which means they grow towards the light. This helps the fungi gain some height for the sporangiophores so that the spores achieve maximum dispersal when they're released.

Sporangiophore production in *Phycomyces* also responds to light. Phycomyces makes two sizes of sporangiophores, giant *macrophores* and dwarf *microphores*. Scientists determined that blue light stimulates the formation of macrophores but inhibits the production of microphores.

FROM POOP TO PRIZES

Phycomyces was first brought into the lab because of a rivalry between two Belgian universities in the 1800s. When a priest from one of the two universities went to Rome to study, he was looking for new subjects to boost the science program at his school so they could better compete with their rivals. He found *Phycomyces* growing on human feces in a cave and took it home to be an experimental subject. For his work describing the growth, development, and behavior of this dung fungus (which he called *Mucor romanus*) he eventually received a royal prize for science.

Phycomyces continued to be a valuable model organism in scientific research, helping scientists understand light responses, metabolic pathways, and gene function. When the scientist Hans Burgeff noticed that *Phycomyces* would only grow in media that contained thiamine (vitamin B1), the fungus became the first organism besides humans known to need a vitamin to grow. After this discovery, *Phycomyces* was used to detect whether thiamine was present in organic materials. Based on his work with the fungus, Burgeff also described and named the process of heterokaryosis. Alfred F. Blakeslee extended our understanding of fungal life cycles by making careful observations of the growth and development of *Phycomyces*, ultimately discovering sexual reproduction in the Mucorales. The Nobel laureate Max

Delbrüff (who won his Nobel prize for his work on DNA in viruses) worked with Blakeslee to figure out how *Phycomyces* responds to light. Delbrüff wanted to understand how tiny signals, such as a few quanta of light or a few molecules of a chemical, could trigger large changes in the growth and behavior of organisms. He created a series of mutant strains of *Phycomyces* that had abnormal sensory responses. These mutant strains were given the prefix names with his initials *mad* in his honor, and they still continue to be useful to scientific investigations today.

Popping your top with Pilobolus

Fungi in the genus *Pilobolus* are sometimes called "hat throwers" because of the unique way they disperse their spores. The sporangium of *Pilobolus* is a flattened dark disk that looks like a hat perched on top of the sporangiophore (see Figure 9-2). When the spores are mature, the sporangiophore acts like a high-pressure water pistol and shoots the hat-like sporangium up into the air. The sporangium has a sticky base that attaches it to plants so it will get eaten by grazing animals. The sporangia pass through the digestive tract of the animals, and the spores get deposited in their poop, which is the fungus's favorite place to grow.

The details for how *Pilobolus* shoots its sporangium are shown in Figure 9-2 (A–D):

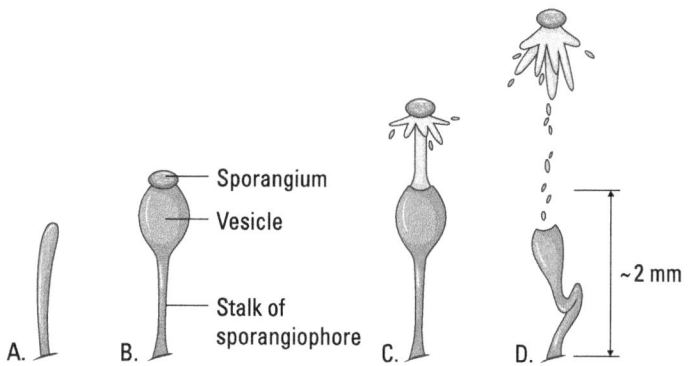

FIGURE 9-2: Spore release in the dung-fungus *Pilobolus*.

A. Sporangiophores of *Pilobolus* are positively phototropic, so they grow upward out of the substrate. The fungus detects light through a ring of light-sensitive pigments located just below the growing sporangiophore.

B. The sporangium forms at the tip of the sporangiophore, which swells to form a balloon-like vesicle called the *subsporangial vesicle*. The subsporangial vesicle acts like a lens, concentrating light back toward the photo-sensitive ring, helping the fungus to keep the sporangium pointed toward the sky.

C. The subsporangial vesicle swells with water and builds up turgor pressure. The vesicle contains dissolved salts and sugar alcohols, which attract water to move into the vesicle.

D. When the pressure inside the vesicle reaches a critical point, a ring of cell wall at its tip breaks, and liquid comes shooting out of the subsporangial vesicle, flinging the sporangium into the air. Scientists say that the average launch speed is 32 kilometers per hour (almost 20 miles per hour) and that *Pilobolus* can shoot its sporangium up to 2.5 meters (about 8 feet).

Exploring the Life Cycle of the Zygomycetous Fungi

In general, zygomycetes frequently reproduce asexually, usually by the production of sporangiospores as shown on the right in Figure 9-3. Asexual reproduction can also occur by *fragmentation*, where pieces of hyphae break off and begin to grow in a new location. A few species in this group can grow as yeast and reproduce asexually by budding. (For more on yeast and budding, head back to Chapter 5.) Sexual reproduction may occur when compatible hyphae encounter each other, resulting in the formation of thick-walled zygospores as shown on the left in Figure 9-3.

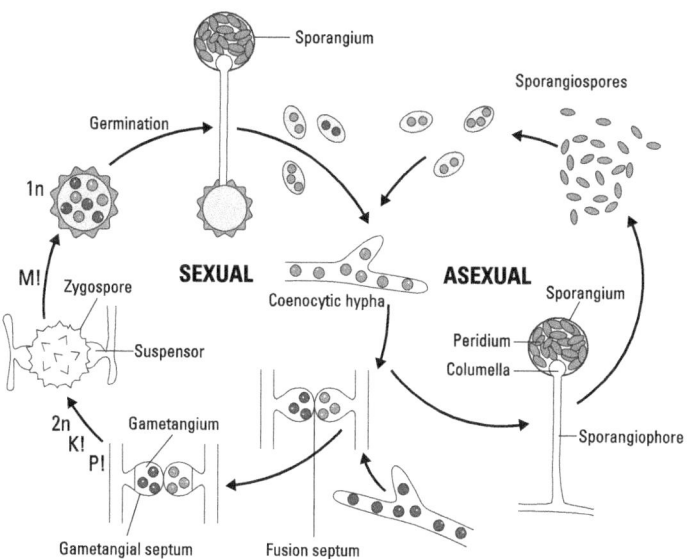

FIGURE 9-3: Life cycle of a zygote-forming fungus (*Mucor* sp.).

Chapter **10**

Morels, Yeasts, Mildews, and More: The Ascomycota

A scomycota is a phylum within Kingdom Fungi. These fungi, called *ascomycetes,* include many familiar fungi such as black and green molds, baker's yeast, powdery mildews, morels, cup fungi, and truffles. In this chapter, you take a look at the characteristics that unite these diverse fungi into one phylum and then explore a few of the important groups within the phylum.

Discovering the Characteristics of the Ascomycota

The ascomycetes, or sac fungi, are the largest group in the fungal kingdom, with over 90,000 species currently identified. Many ascomycetes partner with algae to form lichens (Chapter 3). Some species are pathogenic on plants, while others live on decaying logs and leaves. Some species live only in the dung of certain animals. Some famous ascomycetes shown in Figure 10-1 include the yeasts used to make beer, bread, and wine; the truffles and morels favored in fine dining; and the parasitic species in the genus Ophiocordyceps, which were the inspiration for the fungus in the television show *The Last of Us*.

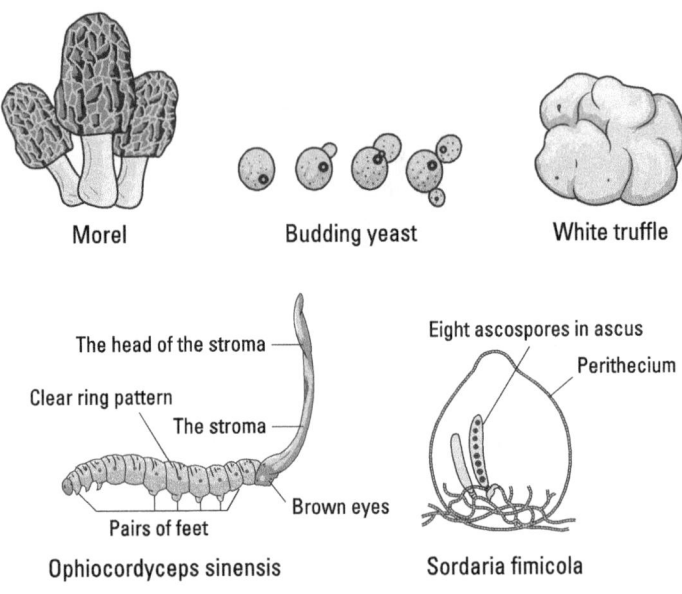

Morel   Budding yeast   White truffle

The head of the stroma

Clear ring pattern

The stroma

Pairs of feet

Brown eyes

Ophiocordyceps sinensis

Eight ascospores in ascus

Perithecium

Sordaria fimicola

FIGURE 10-1: Examples of ascomycetes.

Genetic studies show that the ascomycetes are a well-defined group of related organisms. Most ascomycetes grow by septate hyphae (Chapter 4) that branch often. Small pores in the septae allow cytoplasm, and even nuclei, to move between cells in the mycelium. The individual cells of the hyphae typically have one nucleus, but many species become dikaryotic during sexual reproduction (Chapter 5). Some famous ascomycetes, like the ones that turn fruit juice into wine, don't grow by hyphae at all; instead, they grow as single-celled yeasts throughout their life cycle. During sexual reproduction, all ascomycetes form an ascus.

An *ascus* is a sac-like cell that contains *ascospores* produced by meiosis (Chapter 5).

REMEMBER

WARNING

Before the development of DNA technology, the classification of fungi relied heavily on the appearance of sexual structures. Many ascomycetes (and basidiomycetes, see Chapter 11) look different when they grow asexually versus sexually, and in some species, scientists have never seen the sexual form. This led scientists to create a large group of fungi called the *Deuteromycota* (also known as deuteromycetes or Fungi Imperfecti) that grouped many asexual forms (*anamorphs*, see Chapter 5) together. As scientists discovered more sexual forms (*teleomorphs*, see Chapter 5) and gained the ability to read the genetic code of fungi, they started to sort out this group of fungi, matching asexual with sexual forms and placing organisms in the correct groups based on actual evolutionary relationships. As a result, the old group of Deuteromycota has been officially cancelled. In the cases where scientists still haven't seen a sexual structure, most of the "imperfect fungi" have been assigned to a group based on their DNA sequence.

Examining the different shapes of ascocarps

Once sexual reproduction begins, most ascomycetes produce a fruiting body called an ascocarp, although some do produce naked asci (no ascocarp).

REMEMBER

An *ascocarp* is a multicellular structure that holds the asci. In many ascomycetes, the asci form in a fertile layer called the *hymenium*. In addition to the asci, the hymenium may also contain sterile hair-like projections.

Ascocarps form into several distinctive shapes, as shown in Figure 10-2:

» An *apothecium* is a cup-shaped ascocarp with an exposed hymenium. You can find this type of ascocarp in the cup fungi like the *Peziza* spp. in Figure 10-2, and in the folds on the surface of a morel, which is a mass of apothecia fused into a single structure.

» A *perithecium* is flask-shaped with fungal tissue surrounding the hymenium. An opening called an *ostiole* allows the ascospores to escape when they are mature. These fruiting bodies are small and easy to overlook. They are produced by the red bread mold *Neurospora crassa,* which is an important organism for genetic research, and by several significant plant pathogens like those that cause Dutch elm disease.

>> Some ascomycetes form a structure called a *pseudothecium* that looks very similar to a perithecium. These ascocarps look flask-shaped because they are asci embedded in a depression in the mycelium of the fungus, but they lack an organized ascocarp wall or hymenium.

>> A *cleistothecium* is completely closed and lacks a hymenium. Instead, asci form throughout the structure. Spores are released when the ascocarp breaks open due to weathering or physical force.

>> A *gymnothecium* is a group of asci embedded in a delicate basket of woven hyphae.

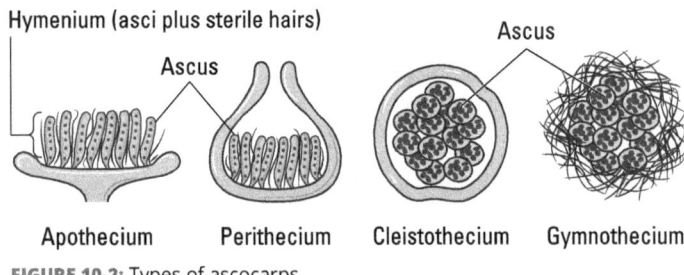

Hymenium (asci plus sterile hairs)

Ascus

Ascus

Apothecium   Perithecium   Cleistothecium   Gymnothecium

FIGURE 10-2: Types of ascocarps.

Observing the colors of molds and mildews

Many ascomycetes produce pigments called *fungal melanins*. These pigments provide additional structural support to the cell wall and protect the fungi from a wide range of potential hazards, including UV damage, drying out, temperature extremes, and free radicals. They're so protective that some scientists refer to them as *fungal armor*. The pigments typically range from brown to black and give color to molds that might otherwise go unnoticed.

TIP

If you've seen greenish-black spots on the caulk in your bathroom, it might be the common mold *Cladosporium*. In some cases, this mold can trigger allergies. A more serious problem is the black mold *Stachybotrys chartarum*, which grows in areas that have significant water damage. This mold makes toxins that can cause respiratory and neurological problems. One visible difference between the two is that *Cladosporium* appears velvety, while *Stachybotrys* looks slimy. Molds love humid environments, so it's a good idea to watch out for musty smells or water stains in your home and be alert for any respiratory symptoms or headaches that coincide with mold growth.

HUNTING FOR TRUFFLES

Truffles have been sending out their powerful lure to humans since ancient times. Beings of mystery, they grow underground to be discovered only by those with the ability to detect their alluring scent — a scent that's rumored to have aphrodisiac properties. The flavor of truffles is described as earthy or nutty, and they are highly prized in culinary circles around the world, so much so that a white truffle (*Tuber magnatum*) weighing 2 pounds, 13 ounces (1.3 kilograms) sold for $330,000 in 2007.

The ancient Greeks debated the nature of truffles, wondering whether they were tuberous roots or something more magical brought forth by Zeus' lightning bolts — after all, they didn't have any stems, leaves, or roots that anyone could see. No one seemed to care enough about their origins to stop eating them, though. Desert truffles (*Terfezia* spp.) are eaten in the Middle East and East Asia, and were probably what the ancient Greeks were eating. Some people think desert truffles may even be the *manna* eaten by the hungry Israelites as described in the Jewish bible. In the 14th century, truffles rose in popularity among the wealthy people of Europe, at least in part due to the strong Italian influence during the Renaissance. Today, Europe remains at the center of truffle appreciation, with two of the most valued truffles being the Périgord black truffle (*Tuber melanosporum*) from Périgueux, France, and the white truffle (*Tuber magnatum*) from Alba, Italy.

It wasn't until the 1700s that scientists figured out that truffles were fungi and not some magical earth children. One early mycologist even spotted developing spores inside of asci in the interior of the truffle. About 100 years later, the Italian physician and mycologist Carlo Vittadini recognized that truffles form associations with plant roots, which scientists now recognize as ectomycorrhizae. This helped people understand why truffles seem to be associated with certain

(continued)

trees, like oak. It turns out that each species of truffle associates with particular trees, forming underground around the tree roots. Modern science also revealed that the reason the ground under these trees is often bare of vegetation is that some truffles produce chemicals that inhibit plant growth. Truffle hunters refer to these bare patches under trees as *terre brulée* (literally burned earth) and regard it as a clue to possible truffle locations.

Truffle hunters know which trees to look under for the truffles that grow in their area, and they usually have help from a truffle-sniffing dog. Originally, European truffle hunters used pigs to sniff out the truffles, but it turns out that pigs are really interested in eating what they find and will damage the fungal mycelium as they dig around. Dogs can be trained to sniff out the delicious morsels and leave them for the human hunter. Good truffle hounds are so highly prized that they are sometimes stolen by unscrupulous truffle hunters. Some truffle hunters locate the fungi without the help of dogs by looking for a yellow fly that's attracted to the smell of ripe truffles. The flies lay their eggs in the soil near truffles so their maggots can use them as food. If a hunter sees these flies around a likely patch of soil under the right tree, they know it's a good place to start digging.

Growing by Yeast and Hyphae with the Taphrinomycotina

The subphylum *Taphrinomycotina* doesn't have a lot of species, relatively speaking, but its members are quite diverse. This group used to be called Archeascomycetes because it's one of the oldest groups of ascomycetes (*archaea*=ancient). Scientists currently recognize six classes in this subphylum, four of which are presented here:

>> The **Taphrinomycetes** is a class of *dimorphic* (*di*=two, *morph*=shape) fungi that grow as both yeast and filamentous forms. This class includes several plant pathogens in the genus *Taphrina* that cause leaf curl and witches' broom diseases in plants. They produce naked asci on the host plant tissue. Inside the asci, the ascospores reproduce yeast cells by budding.

>> The **Pneumocystidomycetes** consists of yeasts in the genus, *Pneumocystis*, which cause lung infections in mammals. In humans, they cause pneumocystic pneumonia.

>> **Schizosaccharomycetes** is the class that includes the fission yeasts in the genus *Schizosaccharomyces*. Unlike the budding yeast shown in Figure 10-1, fission yeast split into two relatively equal cells when they divide by mitosis. Fission yeast (*S. pombe*) is an important organism in genetic research.

>> Class **Neolectomycetes** only contains the genus *Neolecta*. They produce club-shaped fruiting bodies and are one of the fungi commonly called Earth tongues.

Budding with the Saccharomycotina

Most members of the subphylum Saccharomycotina grow as budding yeasts, although some can convert to filamentous growth, and a few grow consistently as filaments. They form their asci within the ascus mother cell and don't form ascocarps. Scientists currently recognize seven classes within this subphylum, the most notable of which is the Saccharomycetes. Two important members of this class are:

>> ***Saccharomyces cerevisiae*** or budding yeast (see Figure 10-1) has been used by people for thousands of years to make beer, wine, and bread. It was the first eukaryote to have its genome completely sequenced, as it is an incredibly important organism for genetic research (Chapter 6).

>> The genus ***Candida*** includes species that can cause human infections (*candidiasis*), such as vaginal yeast infections and oral thrush. *Candida albicans* is dimorphic, sometimes growing as a yeast and other times producing filaments. Under certain environmental conditions, it produces a special type of asexual spore called a *chlamydospore*. These relatively large, thick-walled spores are useful in the identification of this species, but their function is still being investigated.

Forming Ascocarps with the Pezizomycotina

The largest subphylum in the Ascomycota is the Pezizomycotina. This group includes almost all the ascocarp-forming ascomycetes. They are a very diverse subphylum that mostly grow as filamentous fungi, although some species have yeast stages. Almost 40 percent of the fungi in this group form lichens. The other 60 percent have a wide variety of lifestyles, including forming mycorrhizae and being pathogenic on plants, animals, or even other fungi.

Scientists currently recognize 16 classes within this subphylum, a few of which are described here:

>> Almost all the fungi of the **Pezizomycetes** form cup-shaped apothecia like the *Peziza* sp. A notable exception to this is the truffles, which form apothecia that are modified to be completely enclosed (see Figure 10-1) and which form underground. The truffles rely on animals to eat them and disperse their spores in the animal feces (for more on truffles, see the sidebar "Hunting for Truffles" earlier in this chapter). In addition to cup fungi and morels, the Pezizomycetes also includes the morels (*Morchella,* Figure 10-1) and various species commonly known as elfin saddles.

>> The **Leotiomycetes** are filamentous fungi that mostly form cup-shaped apothecia. This class includes *Hymenocyphus ericae,* which forms ericoid mycorrhizae (Chapter 3), and the plant pathogen *Sclerotinia sclerotiorum,* which can attack over 400 species of plants.

>> The **Orbiliomycetes** are mostly saprobic fungi that make small, delicate apothecia. A few, however, have developed a wildly different strategy. Instead of digesting decaying plants, they form lasso-like traps out of their hyphae to catch and digest nematodes (roundworms) in the soil. Scientists are currently researching whether these carnivorous fungi can be used to help control nematode populations that are parasitic on crop plants.

>> The **Dothideomycetes** is an extremely large and diverse class known for producing pseudothecia and bitunicate asci. During sexual reproduction, the mycelium forms little pockets in which the asci develop. The inner wall of the asci swells and then pops up out of the outer wall like a jack-in-the-box to release the spores. Many of the Dothideomycetes are plant pathogens, but others are saprobes, and a few form lichens. *Phaeosphaeria nodorum* causes the disease Septoria nodorum blotch on wheat.

>> Most members of the **Sordariomycetes** form perithecia, like the one shown in Figure 10-2. Most of these fungi live on land, but a few live in the ocean. Many cause diseases in plants, but some cause diseases in animals. Three significant orders within this class are:

- *Hypocreales:* Some of these fungi are saprobes or plant pathogens, while others are parasites on insects or other fungi. The winter worm–summer grass (dōng chóng xià cǎo in Chinese) is a caterpillar that was parasitized by the fungus Ophiocordyceps sinensis (formerly named Cordyceps sinensis). It's been used in traditional Chinese medicine for thousands of years to improve liver function and treat chronic diseases. (For more details on fungi in Chinese medicine, head to Chapter 14, and to hear how *Ophiocordyceps* turns ants into zombies, flip to Chapter 16.)

- *Sordariales:* This order includes fungi that grow on animal dung and decaying plant matter, as well as some plant pathogens, including the species that causes American chestnut blight (see the sidebar "Blighting an American Icon" for details). They are known for producing powerful enzymes that can break down industrial waste, including some plastics. The red bread mold, *Neurospora crassa*, is useful in genetic studies (Chapter 6), as is *Sordaria fimicola* (see Figure 10-1), which grows in the dung of herbivores.

- *Xylariales:* Fungi in this order are known for producing large perithecia that are often dark-colored with thick walls. They often grow on plant litter or animal dung and are important in forest ecosystems because they can break down woody debris. Some species are endophytic, meaning they grow inside plants. Scientists are interested in the chemicals produced by these fungi, some of which have potentially useful effects such as antibacterial, antifungal, and anti-inflammatory activity.

>> The **Eurotiomycetes** includes saprobic and pathogenic fungi, as well as fungi that form lichens. Many members of this class form cleistothecia, but some form perithecia. The order Eurotiales includes many blue and green molds, including the famous *Penicillium chrysogenum*, the source of the antibiotic penicillin (for more on the origins of penicillin, head to Chapter 14). It also includes the genus *Aspergillus*, which can cause infections in humans (aspergillosis) and other mammals.

>> The **Lecanoromycetes** contains most of the fungi that form lichens. Most species form apothecia with asci that have double-layered walls called *bitunicate* asci. The largest order within this class is the Lecanorales with 6,231 species. They produce crustose, foliose, and fruticose lichens. (See Chapter 3 for more on lichens.)

>> The **Geoglossomycetes** are one of the types of fungi commonly called Earth tongues because of the elongated shape of their ascocarps. They're often a dark charcoal color and appear with other fungi in a grassland ecosystem in the United Kingdom known as a wax cap grassland (wax caps are a type of basidiomycete).

BLIGHTING AN AMERICAN ICON

For thousands of years, the American chestnut (*Castanea dentata*) dominated the eastern U.S. forests from Maine to Mississippi. The trees were home to birds and mammals and provided pollen for honey bees. Indigenous Americans valued the chestnuts as a food source and used the leaves as medicine. When European colonists came to the Americas, they incorporated chestnuts into their recipes and found many uses for the tree's rot-resistant lumber.

In the late 1800s, disaster struck in the form of an ascomycete, accidentally brought into America with some chestnut trees imported from Asia. The Asian trees were resistant to the fungus, *Cryphonectria parasitica,* but not the American trees. The fungus spread rapidly in a deadly epidemic that almost wiped the trees from the forests in just 50 years.

The one spark of hope in this story is that the fungus can't quite kill the tree. The fungus causes a canker-type disease called American chestnut blight. It gets into the trees through small openings in the bark, then spreads through the actively growing tissue, forming a ring that "girdles" and kills the growth above the ring. The fungus can't spread downward and kill the roots, however, so the stump of the trees remains alive for some time, periodically sending up shoots in an attempt to grow. Once the shoots begin to mature, however, their bark develops natural furrows that again allow the fungus in. The result is that the once majestic trees are reduced to shrubs trapped in a cycle of rebirth and death.

Scientists have rallied to try to save the American chestnut tree and restore it to its former glory. One group, called the American Chestnut Foundation (tacf.org) is trying several different tactics to make this happen, including traditional breeding efforts that cross American chestnut trees with Asian varieties to try and develop resistant trees, genetic engineering to introduce resistance genes taken from other plant species, and preservation of wild trees in special orchards. Thanks to efforts like these, someday Americans may again enjoy the beauty and benefits of these native trees.

Looking at the Life Cycle of the Ascomycota

The life cycle of filamentous ascomycetes, like the one shown in Figure 10-3, begins when an ascospore germinates to give rise to a growing hypha. As the hypha elongates, it begins to branch so that hyphae spread out through the environment in many directions, forming a complex mycelium. The same mycelium can produce spores asexually or sexually, depending on growth conditions.

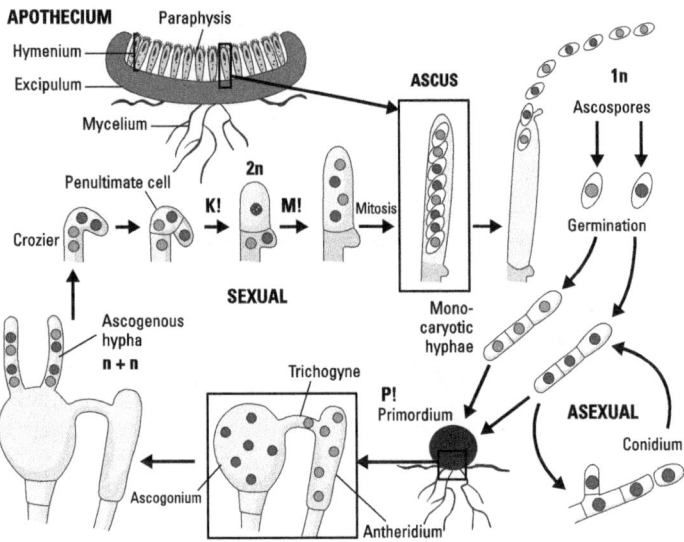

FIGURE 10-3: The life cycle of an ascomycete (*Peziza* sp.).

Exploring the diversity of asexual reproduction

Ascomycetes mostly reproduce asexually, using mitosis (Chapter 5) to produce new cells or spores:

» **Budding yeast,** like baker's yeast (*Saccharomyces cerevisiae*), divide asymmetrically, producing small daughter cells that slowly grow off the mother cell. (For a picture of budding yeast, flip back to Chapter 5.)

» **Fission yeast,** like brewer's yeast (*Schizosaccharomyces pombe*), divide symmetrically by mitosis to produce new cells.

» **Filamentous ascomycetes,** like the one shown in Figure 10-3, produce asexual spores called *conidia* by mitosis. Conidia may be produced from single hyphae, as in Figure 10-3, or hyphae may cluster together to form thicker structures called *conidiomata* that look like small brooms or trees (for more on conidia and to see examples of conidiomata, check out Chapter 4). Some ascomycetes protect their conidiomata under a covering of hyphae.

Conidia are sometimes called "summer spores" because asco-mycetes produce them throughout the growing season. A single mycelium will continue to reproduce as long as food is available, resulting in the release of enormous amounts of conidia.

Reproducing sexually with ascospores

When environmental conditions change, the same mycelia that produced asexual conidia will switch to sexual reproduction. Some ascomycetes are self-compatible and can complete sexual reproduction by themselves; others need to connect with a com-patible strain.

REMEMBER

Homothallic (*homo*=same, *thallus*=plant body) species can complete sexual reproduction by themselves. *Heterothallic* (*hetero*=other) species must connect with a compatible strain.

Hyphae within the mycelium begin to specialize for sexual repro-duction. Some hyphae become *ascogonia,* while others become *antheridia* (see Figure 10-3).

TIP

You can think of ascogonia as the female organ of the fungus because it will create the part that will receive a nucleus. Likewise, antheridia are like the male organ of the fungus because they will ultimately donate nuclei.

A series of events occurs that ultimately leads to the fusion of haploid nuclei, followed by meiosis to produce haploid ascospores within an ascus.

Ascospores form by *free cell formation.* The plasma membrane of the ascus folds inward and wraps around the eight haploid nuclei to form a double membrane around each. Cell wall material is deposited between the two membranes to form the spore walls. Each nucleus receives some cytoplasm, but some remains out-side the spores in the ascus. This *epiplasm* may provide nutritional support to the developing spores.

The shape of the ascus and the method of ascospore release vary among ascomycetes:

>> Asci may be round (*spherical*), club-shaped (*clavate*), or cylindrical. They may have one wall (*unitunicate*) or two walls (*bitunicate*). Some have a very delicate thin wall (*prototuni-cate*) that breaks down easily to release the spores.

>> Some asci open via a hole (*pore*) in the tip or by a small lid (*operculum*) that can pop open. Other asci that don't have an opening have a layer of elastic tissue that can expand and pop the asci open. Still others don't pop open at all, instead releasing spores passively after the outer wall breaks down.

Many ascomycetes have methods to explosively release their spores. For example, the inner wall of those with bitunicate asci absorbs water, swelling and increasing the internal pressure in the ascus until it splits open to shoot the spores out into the air. The narrow, explosive asci of ascomycetes have been nicknamed "spore cannons," and the ascomycetes as a group are sometimes called the "spore shooters." In some ascomycetes, the release of spores from one ascus triggers neighboring asci so that spores come out in a cloud. You can actually hear the sound of the tiny explosions if you get close enough.

Chapter **11**

Portobellos, Puffballs, Stinkhorns, and More: The Basidiomycota

asidiomycota is a phylum within Kingdom Fungi. These fungi, called *basidiomycetes*, include many common culinary mushrooms, puffballs, stinkhorns, bracket fungi, rusts, smuts, jelly fungi, and more. They are a diverse group with varied lifestyles, from free-living saprobes that decay organic matter to plant pathogens, mycorrhizal fungi, and insect symbionts. Many medicinal fungi belong to this phylum, including reishi and lion's mane mushrooms. In this chapter, you take a look at the amazing diversity of these fungi and find out what unites them into one phylum.

What It Takes to Be a Member of the Club

The basidiomycetes, or club fungi, is a large group of incredibly diverse fungi currently estimated to contain over 50,000 species. Like most fungi, they are almost invisible as they grow, spreading their hyphae through their environment. Some of them are saprobes that decay dead organisms, while others attack living plants. Many species form beneficial mycorrhizae with living plants, and a few form lichens. (For more on these relationships, head back to Chapter 3.) Basidiomycetes also form beneficial and parasitic associations with insects.

When basidiomycetes reproduce sexually, they form fruiting structures called *basidiocarps* that are larger and more visible than their mycelia. You would probably recognize quite a few of them as fungi you've either eaten or seen in the market, such as button mushrooms, chanterelles, oyster mushrooms, and wood ear fungi. In addition to these familiar shapes, basidiomycetes can appear as thin crusts, bracket fungi, round puffballs, pointy stinkhorns, or charming bird's nest fungi. (See Figure 11-1 for some examples.) All of these different shapes of basidiocarps can be organized into a few overall growth patterns:

» **Resupinate** fungi grow like crusts with one side of the mycelium attached to the surface.

» **Pileate** fungi have distinct caps (*pilei*). Pileate fungi may also have a stalk (*stipe*). The mycelia of bracket fungi produce both a shelf-like cap and a resupinate portion that attaches the cap to the substrate. This type of basidiocarp is also referred to as *effused-reflexed*. The reishi shown in Figure 11-1 is an example of a bracket fungus.

» **Coralloid or clavarioid** fungi produce erect tubular basidiocarps that may branch so that the basidiocarps look like corals that grow in the ocean. (See the coral fungus in Figure 11-1.)

The basidiomycetes are so diverse that you might wonder how scientists decided that they're all related to each other. The feature that unifies this phylum is that, when these fungi reproduce sexually, they all support their sexual spores on a structure called a basidium. DNA analysis also shows that the basidiomycetes are a cohesive group whose closest relatives are the ascomycetes (Chapter 10).

A *basidium* is a cell on which basidiomycetes produce their sexual spores, called *basidiospores*. In many basidiomycetes, the basidium is club-shaped, which is why these fungi are nicknamed the club fungi.

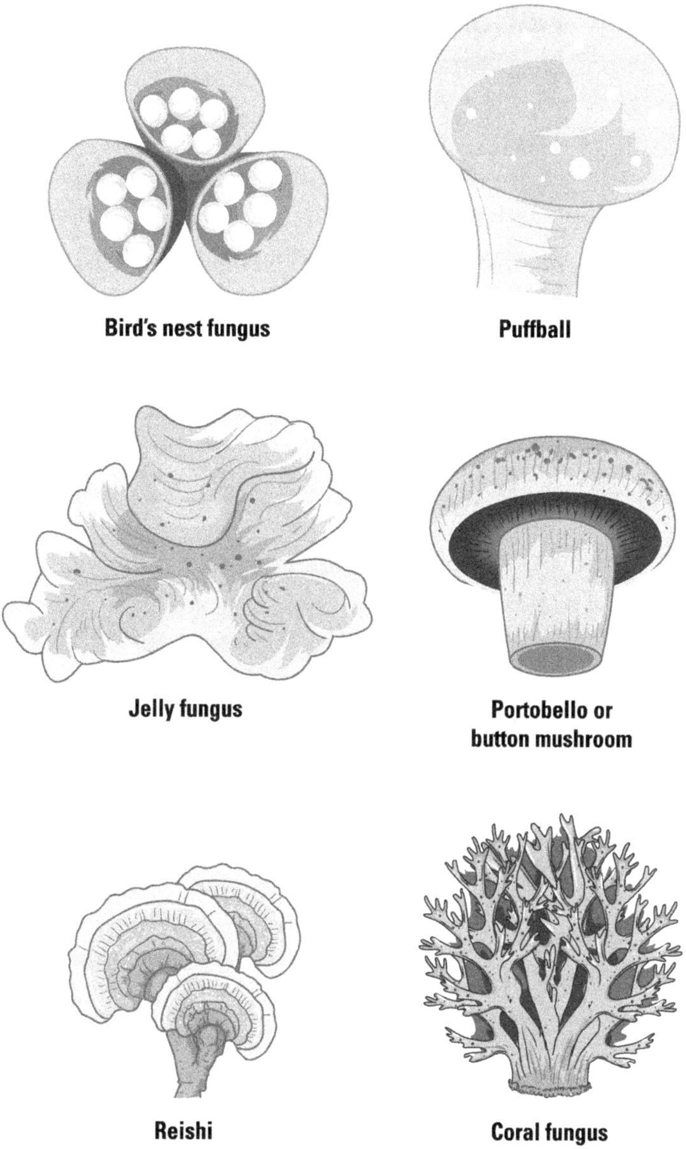

Bird's nest fungus

Puffball

Jelly fungus

Portobello or button mushroom

Reishi

Coral fungus

FIGURE 11-1: Examples of basidiomycetes.

Basis is the Greek root meaning base, so a basidium is a supportive base or pedestal for the spores.

TIP

Exploring the life cycle of the basidiomycetes

The life cycles of basidiomycetes vary between different groups. Some basidiomycetes spend most of their time as yeast, while others alternate between yeast and filamentous forms. Most mushrooms, on the other hand, only grow as filaments.

Dimorphic basidiomycetes alternate between a yeast form and a filamentous form, while *monomorphic* species only seem to grow in one form or the other (*di*=two, *mono*=one, *morph*=shape).

Some basidiomycetes favor sexual reproduction, but asexual reproduction occurs by various methods in different species:

>> **Budding:** Some basidiomycetes grow as single-celled yeasts. These cells can reproduce by mitosis. (For details on mitosis, head back to Chapter 5. For a quick look at budding in the ascomycetes, go to Chapter 10.)

>> **Fragmentation:** Pieces of mycelia can break away, settle on new substrate, and begin to grow. When single cells break away from hyphae, scientists call them *arthrospores*.

>> **Asexual spores:** Some basidiomycetes produce asexual spores called *conidia* by mitosis. Other species form short hyphal branches called *oidiophores* that produce a series of spores called *oidia* at their tips.

Despite these differences, some things are generally true of basidiomycete life cycles, as shown in Figure 11-2:

>> Unlike other fungal groups that predominantly grow as haploid cells, basidiomycetes spend most of their time as dikaryotic cells that contain two haploid nuclei. These dikaryotic cells result from the cytoplasm fusion, or *plasmogamy*, of two hyphae.

>> Like other fungal groups, nuclear fusion, or *karyogamy*, is almost immediately followed by meiosis. (For details on meiosis, head back to Chapter 5.)

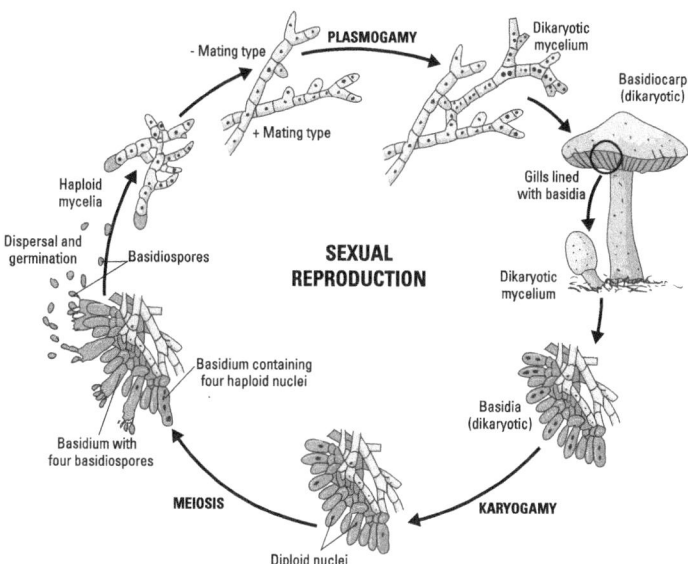

FIGURE 11-2: The basidiomycete life cycle.

Catapulting spores

Basidiomycetes are serious about spore dispersal, flinging their mature spores outward with incredible force and speed. The fungi fling their spores with a *surface tension catapult* that uses the movement of a water droplet to generate force. Figure 11-3 illustrates how this catapult works:

1. The sterigma holds the basidiospores to the basidium.

2. Condensation causes water droplets to form on the spore:

   a. One drop forms on a projection on the spore called the *hilar appendix*. Scientists call this droplet the *hilar droplet* or the *Buller's drop* after the scientist who first described the surface tension catapult.

   b. Another drop forms on the side of the spore. Scientists call this drop the *adaxial drop*.

3. Initially, the two drops remain separated by a small barrier of water-repellent material on the spore, but eventually, Buller's drop gets large enough to touch the adaxial drop.

4. When the drops touch, they rapidly merge into one big drop that centers on the side of the spore.

5. The sudden movement of water transfers mass onto the side of the spore in the direction away from the basidium, causing the spore to detach from the sterigmata and go catapulting into the environment. Because the spores are shot out with force, scientists refer to them as *ballistospores.*

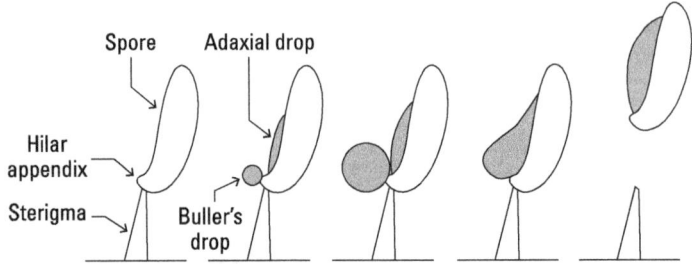

FIGURE 11-3: A surface tension catapult.

Many basidiomycetes use surface tension catapults to release their spores, although a few types, such as puffballs and bird's nest fungi, have lost this ability:

>> Bird's nest fungi like the one shown in Figure 11-1 use raindrops and passing animals to release their spores. The structures that look like eggs in the nest are actually sacs of spores called *peridioles.* Raindrops splash into the cup-shaped *peridium* of the fungus, knocking the "eggs" out of the "nest." When the peridioles launch, some have a long, sticky tail of hyphae called a *funiculus.* This sticky tail attaches to nearby plants so that the peridioles dangle in the air. When an animal passes by or eats the plant, it takes the fungal spores with it, dispersing them to new environments.

>> In puffballs (see Figure 11-1), the basidia are completely enclosed in a sphere of protective tissue called the peridium. In some puffballs, the peridium flakes away as the puffballs mature, allowing spores to escape into the air. In other puffballs, the outer peridium flakes away, leaving behind a thin inner layer that has a hole in the top. When raindrops hit the puffball, clouds of spores shoot into the air. A passing animal's foot can also cause a geyser of spores, as anyone who has ever stomped a puffball can tell you.

Recognizing Familiar Fungi in the Agaricomycotina

Many of the mushrooms you're familiar with are probably in the subphylum Agaricomycotina, which contains approximately two-thirds of all known basidiomycetes. Some of them form basidia that don't have any septa, called *holobasidia*, while others form basidia that do have septa, called *phragmobasidia*.

Scientists divide the Agaricomycotina into three classes:

>> The **Tremellomycetes** grow in many forms, including yeasts, species that produce gelatinous basidiocarps, and lichen-forming species. Species that produce gelatinous basidiocarps are commonly referred to as jelly fungi. These basidiocarps range from thin layers of mycelium to wrinkly cushions to folded, leafy structures like that of the snow ear (*Tremella fuciformis)* shown in Figure 11-1. Many species in this class are dimorphic, growing as yeast for part of their life cycle. Most species grow as saprobes, but some are parasitic on plants or insects. When water is scarce, jelly fungi can become dry and brittle, but after rain, they become gelatinous again.

>> The **Dacrymycetes** are a small group of wood-decaying fungi with gelatinous and brightly colored basidiocarps. They're often bright yellow or orange due to the presence of *carotenoids,* which are the same pigments that give carrots their color. They can often be seen growing on dead branches and tree stumps in wet weather. They have very distinct forked basidia that resemble the letter Y.

>> The **Agaricomycetes** are the largest and most diverse class in this subphylum. Most species are saprobes, including wood-decaying fungi that cause brown and white rot. (For more details on how fungi break down wood, see Chapter 2.) Other species play a wide variety of roles in nature, including plant pathogens, ectomycorrhizae, insect symbionts, and lichens. Some species even trap and devour nematodes (roundworms) or amoebas.

Delving into the diversity of Agaricomycetes

The class Agaricomycetes includes gilled mushrooms, polypores, boletes, chanterelles, bracket fungi (shelf fungi or conks), puffballs, coral fungi, and false truffles. This class is so large and diverse that it's impossible to do it justice in just a few pages, so I've selected a few representatives within each order to highlight some of the diverse features of this class. To learn more about these amazing fungi, check out some of the resources in Chapter 18:

>> The order **Agaricales** includes most of the mushrooms that have caps, stalks, and gills, such as the edible button or portabella mushrooms (*Agaricus bisporus,* shown in Figure 11-1) and oyster mushrooms. In *agaricoid* fungi like these, the hymenium lines the gills, producing the basidia under the protection of the cap. The beautiful but often toxic *Amanita* species, including the iconic fly agaric with the white-spotted red cap, are in this order (the white spots are actually warts, for more on *Amanita* see Chapter 13), as is *Armillaria ostoyae,* the fungus that wins the title for largest organism in the world (Chapter 16).

Some members of the Agaricales form very different types of basidiocarps. The elegant coral fungi, such as the *Clavulina* sp. shown in Figure 11-1, produce clavarioid basidiocarps with hymenia that cover the entire surface. Several puffballs, including the giant puffball shown in Figure 11-1, are also part of this group. In these *gasteroid* fungi, the hymenium is completely enclosed within the basidiocarp.

>> The boletes of order **Boletales** are highly prized among mushroom hunters and include the delicious porcini mushrooms (also known as king boletes). These fungi have a cap and a stalk, but instead of gills, they have pores on the underside of the cap. The hymenium lines the pores where it forms the basidia.

>> The **Gloeophyllales** are tough bracket fungi that degrade conifer wood. They range in color from dark brown to gray, often have concentric bands of color on their surface, and may be hairy or shaggy. They produce basidia on their lower surfaces in pores or in maze-like gills.

>> The fungi grouped together in the **Hymenochaetales** based on their DNA sequence produce a wide variety of basidiocarps. Some grow as thin crusts on the underside of branches, others form brackets or coral-like shapes, and still others grow as classic stalked mushrooms. Most are saprobes, but species in the genera *Inonotus* and *Phellinus* attack living trees, causing significant losses in the forestry industry. The mycelium of *Inonotus obliquus*, known as chaga, looks like a burn scar as it grows on trees like birch. The people of Russia and northern Europe have used chaga for centuries to make a tea that is believed to boost immunity. Western scientists are currently testing chaga in the laboratory for a wide variety of impacts, including its ability to reduce inflammation, fight cancer, and lower blood sugar.

>> The **Polyporales** contains many bracket fungi that have textures ranging from fleshy to hard and woody. A few species grow as crusts or form stalked mushrooms. The fungi are called polypores because they produce their basidia in many small holes rather than on gills (*poly*=many, *pore*=hole). Many fungi in this group cause wood rot (Chapter 2). *Ganoderma sichuanense*, known as lingzhi in China and reishi in Japan (see Figure 11-1), is used in traditional Chinese medicine to promote health and long life. In the wild, it grows as a shiny bracket fungus that appears at the base of trees like maples. Modern research supports the medicinal value of this mushroom in fighting cancer and supporting the immune system. (For more on medicinal uses of fungi, head to Chapter 14.)

>> The **Russulales** is a large group, the majority of which grow as ectomyccorrhizae and produce gilled mushrooms. Some species grow as crusts, polypore bracket fungi, and coral fungi. Members of the genus *Russula* are sometimes called brittle gills because their gills can't be bent without breaking. This feature, combined with their brightly colored caps, makes them relatively easy to identify. Many species are edible, although some produce toxic compounds. Mushrooms in the genus *Lactarius* are called milk-caps because of the fluid they release when cut or damaged, which is usually white or cream colored. This milky fluid is called latex.

>> Fungi in the **Thelephorales** grow in a wide variety of forms, from crust-like to coral fungi to polypore bracket fungi. Some species are *hydnoid*, meaning they produce their basidia on spines or tooth-like projections. Hydnoid fungi often look shaggy because of the projections that hang off their basidiocarps.

WARNING

Fungi that look alike aren't always closely related. Just like birds and bats both have wings but aren't very closely related, some fungi that aren't very closely related have similar characteristics. Puffballs and other fungi with closed basidiocarps used to be grouped together in the Gasteromycetes. Likewise, basidiomycetes with hymenia that were exposed on their gills or pores used to be grouped together in the Hymenomycetes. DNA science revealed that these characteristics developed separately in unrelated groups of fungi, so these old categories are no longer valid.

Interacting with insects

Basidiomycetes form relationships with insects that span the range from mutually beneficial to predatory. Fungi are an important food source for some insects, so much so that termites and ants farm their favorite species. (For details on ants that farm fungi, head to Chapter 16.) Scientists have documented entire communities of fungi-loving beetles that live in and feed on mushrooms, helping to disperse the spores by their activity. As fungi live in close association with insects over long periods of time, both the fungi and the insects may develop special structures or behaviors that support the relationship.

Pretending to be a termite

One slightly mysterious association is between termites and fungi that pretend to be termite eggs. Japanese scientists discovered this relationship when they found small round balls mixed in with termite eggs inside termite nests. When they cut the balls open, they discovered that they were made of mycelia. Further analysis revealed that these termite balls were sclerotia of a basidiomycete.

REMEMBER

Sclerotia are balls of fungal mycelium with thick cell walls and slow metabolic rates. They contain food reserves and can survive harsh environmental conditions.

The termite balls are slightly smaller than the actual termite eggs, but the blind termites who tend the eggs don't seem to notice. The termite balls are smooth and covered with a chemical surface similar to that of the eggs, so the workers pick them up and groom them just as they would an egg. As long as the termite balls get groomed, they don't begin to grow. When the termites go underground for the winter, however, the fungus grows and parasitizes any eggs that were left behind.

On the surface, this relationship seems advantageous to the fungus. It receives care in a protected environment and then later may get to eat termite eggs. But scientists still have many questions. Termite balls are found in the majority of colonies of the termite *Reticulotermes*, but scientists don't know if the fungus gets transferred between nests by termites. They're also trying to figure out if this lifestyle gives the fungi a competitive advantage over other fungi in their environment. And, until recently, they hadn't seen spore production in the fungus and weren't exactly sure how to identify it.

The discovery of termite balls occurred in Japan, so it seems fitting that Japanese scientists recently revealed the identity of the fungus that makes the balls. The anamorph, or asexual form, of the fungus was named *Fibularhizoctonia*. When scientists read the DNA extracted from the balls, the code suggested that this fungus was a member of the genus *Athelia*, many of which grow as saprobes on rotting wood. Scientists collected samples of free-living *Athelia* from around Japan. Next, they got the fungi to grow in the lab so they could observe their life cycles. One of the species of fungi produced sclerotia in the lab that were identical to those found in the nests, as confirmed by physical appearance and DNA sequence. Based on their observations, the scientists proposed that this newly identified fungus be named *Athelia termitophila* (*termitophila*=loves termites).

Hunting prey for dinner

Most species of fungi get their dinner by decomposing the dead, but there are a few less patient species that go out and catch their dinner while it's still wiggling. *Hohenbuehelia* is a slightly gelatinous bracket fungus that decays wood. It gets all the carbon it needs from wood, but not enough nitrogen to build its proteins. So, as it grows through the wood, it produces sticky knobs in its hyphae. When nematodes (roundworms) come too close, they get stuck on the knobs. As they thrash around, they hit other knobs until they have no hope of escape. The fungal hyphae grow into the worm and digest it for its protein and nitrogen content.

The edible oyster mushroom *Pleurotus ostreatus* and many of its close relatives use chemical warfare to hunt nematodes. These fungi produce little bags called *toxocysts* on their hyphae that contain a poison. When nematodes brush the bags, the bags pop and release the toxin, causing paralysis and killing the nematodes. The fungal hyphae invade the corpses and digest the worms.

Partnering with wood-boring wasps

The wood-boring wasp *Sirex noctilio* uses a white rot fungus, *Amylostereum areolatum,* to make sure their larvae get a good start in life. The fungus is so important to the successful reproduction of the wasp that the wasp carries the fungus in a special organ called a *mycangium*. When the wasp drills its ovipositor into a pine tree, it deposits mucus and fungi along with its eggs.

The fungus digests the wood of the tree, helping the developing wasps as they grow and change. At first, the fungal digestion dries the wood slightly, creating a better environment for the wasp eggs. After the eggs hatch and the larvae begin to grow, they feed on both the fungus and the predigested wood. As the larvae transition to the adult wasp form, they move some of the fungus into their own mycangia so the cycle can begin again.

The mutualism between the wasp and the fungus gives both species a boost to their success. Insects don't produce the enzymes to digest wood, so the fungus makes it possible for the wasp to access a new habitat and food for its offspring. Not only do the wasps carry the fungus to new hosts, but they also secrete a phytotoxic venom in the mucus they deposit along with their eggs. The phytotoxic venom damages the plant tissue (*phyto*=plant, *toxin*=poison), lowering the ability of the plant to defend itself against the fungus.

Unfortunately, what is good news for the wasp and the fungus is bad news for the trees. *S. noctilio* is native to Europe, Asia, and North Africa. In those parts of the world, it causes minor damage to forests. But when the wasp hitched a ride in wooden packaging to other countries, the impact on trees was much greater. Since the wasp was introduced to the United States in 2004, it's spread throughout the northeastern states and into Canada. The wasp causes the most damage when it first moves into a new forest and when trees are crowded or weak. It has caused millions of dollars of damage to the lumber industry in countries like Australia and is projected to do so in the United States.

Generating Rusts on Plants with the Pucciniomycotina

The subphylum Pucciniomycotina consists of about 8,400 species, most of which are plant pathogens called *rusts* because of the rusty red color of the lesions they produce on plants. A few pucciniomycetes form mycorrhizae, and some are free-living yeasts. Others are insect pathogens or parasites of other fungi.

The plant pathogens in this group cause significant crop damage. The fungus disrupts the plant tissue, interfering with photosynthesis and reducing plant growth and fruit production. In 2024, it was estimated that the global economic impact of rust diseases was a loss of one billion dollars from wheat crops and three billion dollars from coffee. In countries where fungicides aren't available, rusts can reduce cereal crops by as much as 75 percent and legume (bean) crops by 50 percent.

The relationship between a pathogenic rust and the plant it infects is very specific.

Host range is the range of species that a pathogen can infect. Plant pathogenic rusts have narrow host ranges.

Many rust species have specialized to become very good at infecting their host species, but the trade-off is that they often lose the ability to infect a wide range of hosts. As different rusts evolved along with their hosts and became specialized, they often changed enough to be considered new species, which led to a great deal of diversity within this fungal group.

Rusts have some of the most complex life cycles of any organism on Earth. Some of them require two hosts to complete asexual and sexual reproduction, and they can make as many as five different types of spores.

Heteroecious species require two hosts to complete their life cycle. *Autoecious* species require just one host.

Puccinia graminis, which causes black stem rust on wheat, is a heteroecious rust. Wheat (*Triticum aestivum*) is the primary host of the rust fungus because the fungus can infect wheat and reproduce indefinitely via asexual reproduction. Barberry (*Berberis vulgaris*) is the alternative host for the fungus because it must infect this plant to reproduce sexually.

Over time, plant pathogens have become very good at what they do, evolving specialized structures and proteins that allow them to invade and take advantage of their host plant. Rusts show several examples of these strategies.

Dikaryotic spores called *urediniospores* germinate to produce hyphae that grow over the wheat plant surface. When the hyphae detect an opening called a *stomate* on the surface of the plant, they form specialized appressoria that help them invade the plant tissue. The hyphae seem to actually be able to detect the topography of the leaf surface to locate the stomates.

An *appressorium* is a specialized hyphal structure that allows germinating spores to adhere to host tissues.

When hyphae growing through the wheat plant encounter photosynthetic cells called mesophyll cells, they form absorptive haustoria.

A *haustorium* is an absorptive structure inserted into a living cell. In the case of rusts, haustoria are specialized hyphal structures that the fungus inserts into the plant cell. The cell wall on the haustorium is very thin, and it's tightly pressed to the host cell membrane so the fungus can absorb nutrients.

When some rusts enter the sexual phase of their life cycle and produce spermatia, they reprogram plants to produce fake flowers called *pseudoflowers*. These pseudoflowers look attractive to insects and even produce enticing smells and nectar. When insects visit the pseudoflowers, they pick up spermatia from the fungus which they then carry with them as they visit other plants, facilitating sexual reproduction of the fungus.

Forming Smuts on Plants with the Ustilaginomycotina

The subphylum Ustilaginomycotina consists of about 1,700 species, most of which are plant pathogens called *smuts*. The smuts are usually dimorphic species that grow as yeast during their asexual stages and then transition to dikaryotic mycelia when they infect plants. The smuts mainly infect grasses such as barley, wheat, and sugarcane, but they can also infect potatoes. They get their name from the masses of black resistant spores called teliospores that look like soot or smut on the plants. A few ustilaginomycetes grow as free-living yeasts or animal pathogens.

The plant pathogens in this subphylum have two phases to their life cycle:

» In the *saprobic phase*, the fungi live as haploid yeast on decaying matter and reproduce by mitosis. If two compatible yeast cells fuse, they create a dikaryotic infection hypha that can infect a host plant.

» In the *phytoparasitic phase*, the dikaryotic mycelium grows through plant tissue. The hyphae mostly grow between the plant cells, but sometimes form haustoria inside the plant cells. This phase continues until the formation and release of teliospores.

When *Mycosarcoma maydis* infects corn plants (*Zea mays*), its mycelium spreads throughout the plants. As it forms masses of mycelia, it triggers the formation of plant tumors, or galls, in the leaves, stems, and flowers of the plant.

REMEMBER

Plant *galls* are abnormal growths on plant tissues. Galls are typically triggered by injury, such as that resulting from the activity of insects and fungi.

The galls that form on the leaves and stems of corn plants typically remain small, but the ones that replace kernels on the corn ears swell, become pale in color, then turn gray, as shown in Figure 11-4. Eventually, they burst open to reveal the masses of black teliospores within.

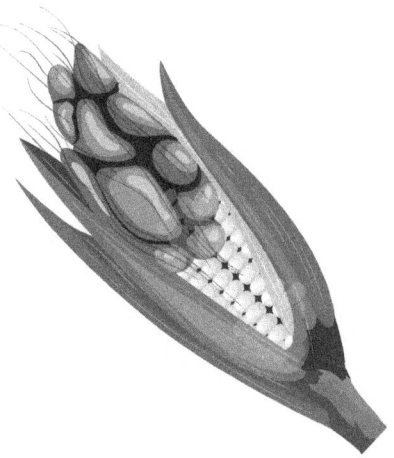

FIGURE 11-4: Huitlacoche or corn smut caused by *Mycosarcoma maydis* infection of *Zea mays*.

For some farmers, corn smut represents a problem because it lowers the yield of ears that can be sold. This isn't true in Mexico, however, where corn smut is prized as a delicacy and referred to as huitlacoche, the name given to the corn smut by the Aztecs. The Aztecs developed sophisticated farming systems to grow crops like tomatoes, beans, peppers, squash, and corn (maize). At some point, corn smut infected their corn crops, but instead of throwing the food away, they tried it. And liked it.

Today, some people refer to huitlacoche as Mexican corn truffles. They say the flavor is earthy, smoky, and a little bit sweet. Beyond its flavor, it has excellent nutritional value because of its high protein content. It even contains essential amino acids that are normally missing from corn and produces compounds that have anti-tumor activity. The fact that huitlacoche tastes good and is good for you hasn't gone unnoticed by the rest of the world. Some farmers now grow corn and deliberately infect it with fungal spores to produce this gourmet fungus.

3

Putting Fungi to Use

Take a look at the nutritional benefits of fungi and fungi's role in creating a variety of fermented foods.

Explore how some cultures have a long history of using hallucinogenic mushrooms for spiritual purposes.

Discover the healing properties of fungi and how people have been using fungi as medicine for thousands of years.

Investigate the different ways to farm mushrooms as well as the best methods for growing mushrooms at home.

Chapter **12**

Fine Dining with Fungi

ungi don't always get the recognition they deserve for their importance, not just in adding flavor to meals, but also for the health benefits they provide people and the planet. Mushrooms contribute a savory impact to foods while also supplying beneficial fiber that improves gut health. Yeast and certain molds create fermented foods, from bread to miso, which serve as dietary staples for some groups of people. In this chapter, you explore the nutritional benefits of fungi, both domesticated and wild, and then take a look at fungi's role in different types of fermentation.

Feeding Ourselves with Fungi

Many people appreciate the flavors of fungi, including them in everyday dishes like soups and stir fry, and in gourmet meals like mushroom risotto or duck breast with a morel mushroom cream. Fungi provide a savory kick to food and are increasingly being used instead of meat as a main-dish ingredient.

In addition to their flavor possibilities, fungi are nutritional powerhouses that provide protein, vitamins, and minerals, while also being an excellent source of dietary fiber. And if that weren't enough, fungi also benefit health with their antioxidant, antimicrobial, anti-inflammatory, anticancer, antitumor, and immunostimulatory effects. Because of their many health-promoting features, fungi can be considered functional foods.

A *functional food* is a food that provides a benefit beyond its nutritional value.

Nutritional recommendations often encourage people to eat more fruits and vegetables as a way to increase their intake of fiber, vitamins, and antioxidants, while lowering their intake of saturated fats from animal products. In these recommendations, mushrooms are usually included in the list of vegetables as a single entry among a long list of plant foods. Likewise, in artwork that promotes healthy diets, a small clump of button mushrooms is usually drawn in among a huge crowd of plant-based vegetables (or sometimes, they're not included at all). As information about the health benefits of mushrooms continues to grow, more people are realizing that mushrooms should be more than just a sidenote in nutritional recommendations.

Benefitting from diverse fibers

Many countries around the world are experiencing epidemic levels of obesity, accompanied by increased levels of chronic diseases such as heart disease, diabetes, and colon cancer. One of the major risk factors for developing chronic diseases such as these is a diet low in fiber intake.

Dietary fiber consists of complex carbohydrates that can't be digested by humans. They pass through the stomach and small intestine, then into the large intestine (colon), where they can be digested by gut microbiota (see Figure 12-1).

Nutritionists categorize dietary fiber into two categories based on whether it dissolves in water. Mushrooms contain both types of dietary fiber:

>> **Insoluble fiber** doesn't dissolve in water. It adds bulk and helps hold water in stool, making it easier for it to pass through the digestive system.

>> **Soluble fiber** dissolves in water, forming a gel-like substance in the digestive system. It helps slow digestion and keeps you feeling full longer. Studies show that soluble fiber lowers the levels of bad (LDL) cholesterol and glucose in the blood, which may help prevent heart disease and diabetes.

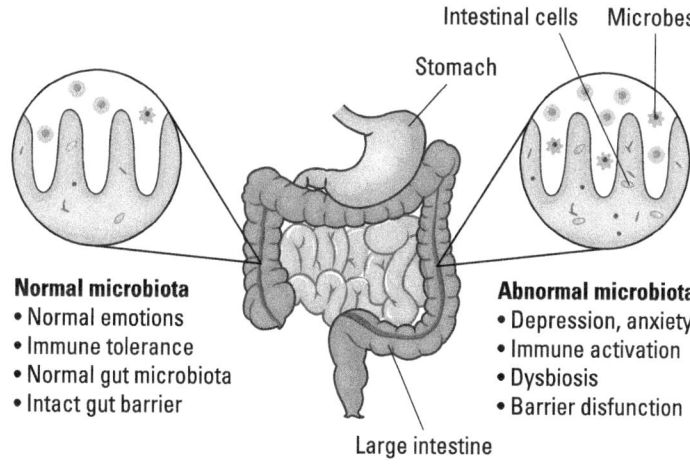

Normal microbiota
- Normal emotions
- Immune tolerance
- Normal gut microbiota
- Intact gut barrier

Abnormal microbiota
- Depression, anxiety
- Immune activation
- Dysbiosis
- Barrier disfunction

FIGURE 12-1: Gut microbiota in the human digestive system.

Dietary fiber isn't just good for your health; it also benefits the microbes that live in your gut.

REMEMBER

Your *gut microbiota* are the microbes that live in your intestinal tract, mostly in your large intestine.

You may already know that you have more microbes living on your body than you have cells of your own and that these microbes are important to your health. Scientists estimate that for every cell in your body, you have between three and ten microbial cells living on you. And 95 percent of these live in your intestinal tract, so the health of your gut microbiota can have a large impact on your overall health. Scientists have correlated the composition of the human gut microbiota with diseases such as obesity, irritable bowel syndrome (IBS), liver disease, and cancer. Recently, scientists have even shown that changes in the types of microbes in your gut can lead to changes in mood.

REMEMBER

Dysbiosis refers to a shift in your microbiota that contributes to poor health (*dys*=bad, *bios*=life).

Although scientists are still researching which species of microbes are associated with a healthy gut microbiome, it seems that a more diverse microbial community is associated with better health. Factors such as age, environment, and drug therapy can affect your gut microbiota, but the biggest impact may be caused by what you eat — especially how much fiber you eat. As dietary fiber intake decreases, so does the diversity of the gut microbiota.

TIP

To have a healthy and diverse microbiota, you need to eat a wide variety of dietary fiber.

Not only are fungi a good source of dietary fiber, but they also provide fiber not found in fruits and vegetables. Fungi aren't plants, and the complex carbohydrates that make up the fungal cell wall are different from those found in plant cell walls (for a refresher on the fungal cell wall, flip back to Chapter 3). The dietary fiber found in fungi includes chitin, hemicellulose, and beta-glucan (β-glucan). These fibers act as prebiotics for the gut microbiota.

REMEMBER

A *prebiotic* is something that feeds your gut microbiota. (In contrast, a *probiotic* is a food that contains live microbes that you are trying to add to your microbiota.)

By eating many different types of fiber, you can encourage diversity and health in your microbiota. Only microbes that have the right enzymes can break down any one particular fiber, so when you offer diverse types of fiber, you allow more microbes to take part in the feast. Populations of beneficial bacteria grow and inhibit the growth of pathogenic bacteria, either by releasing inhibitory molecules or by changing growth conditions such as acidity.

When your gut microbiota breaks down long dietary fiber for food, it produces small molecules called short-chain fatty acids (SCFAs) such as acetate, propionate, and butyrate. SCFAs have many beneficial effects on your health. For example,

>> **SCFAs provide food for the cells that line your large intestine,** giving them the energy they need for reproduction and repair.

>> **SCFAs affect digestion and hunger** by stimulating the large intestine to make hormones such as glucagon-like peptide-1 (GLP-1) and peptide tyrosine tyrosine (PYY). GLP-1 helps lower sugar in the blood by promoting the production of insulin and increasing the body's sensitivity to insulin. PYY can reduce food intake by slowing gastric emptying and helping create a feeling of satiety.

>> **SCFAs regulate the immune system** by stimulating the production of cells and molecules in your immune system

that identify and destroy potential pathogens. They also inhibit the production of factors that trigger inflammation.

>> **SCFAs may help prevent cancer** by signaling cancer cells to stop dividing. They may also trigger cancer cells to kill themselves.

Mushrooms rich in beta-glucans may increase SCFA production, especially in older people.

Tasting fabulous flavors

If you are only familiar with a few common varieties of mushrooms, it may surprise you to learn that mushroom flavor can range from earthy to fruity, nutty, fishy, and even slightly sweet. The taste can also vary depending on whether you eat fresh or dried mushrooms. Many people like the texture, or mouthfeel, that mushrooms give to a dish as well as the satisfying note of a rich, savory taste.

Umami is the taste people recognize as savory. It's a taste response to the presence of certain amino acids found in proteins, especially the amino acid glutamate.

Umami is one of the five basic tastes recognized by human taste buds (see Figure 12-2). The others are sweet, salty, sour, and bitter. (Some evidence suggests that humans can also taste fats, but that is still being investigated.)

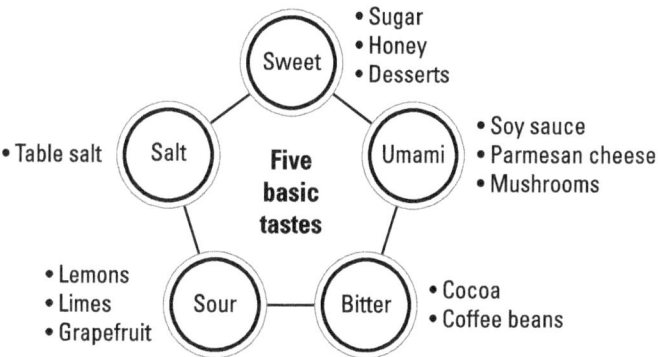

FIGURE 12-2: The five basic tastes.

WARNING

The common diagram of the human tongue that shows the basic tastes mapped to distinct regions of the tongue is wrong. In reality, people have taste receptors for all the basic tastes on several parts of the tongue, the roof of the mouth, and even the throat. The familiar but faulty map grew out of an artistic rather than precise presentation of some taste experiments done around 1901. Since then, many scientists have shown that the map isn't accurate, but sometimes it's just hard to get rid of a wrong idea that's taken root in the public consciousness. The correct map would show that certain regions of the tongue, like the tip and the sides, are more sensitive to taste overall, but that these regions can detect all of the basic tastes.

Even if they don't know its scientific name, most people are familiar with the mushroom, *Agaricus bisporus*. This species is sold in three different forms in many grocery stores. When this mushroom is young, it's small and white and is sold in stores as button mushrooms or champignon (see Figure 12-3). As it matures, it becomes darker in color and slightly more firm and is often sold under the label of cremini or baby bella mushrooms. Allowed to reach its full size, it becomes the giant portobello mushroom that can easily reach six inches across. These mushrooms are firm and relatively sturdy, so they hold up well when cooked in soups or sautéed. The large size of portobellos makes them a favorite for grilling.

Grocery stores and farmers' markets may occasionally offer you new and interesting mushrooms to try. Some of these are domesticated varieties that are grown commercially. Occasionally, you may even have the opportunity to purchase wild-harvested mushrooms. Each type of mushroom has its own unique flavor and texture and lends itself to particular uses in the kitchen. There are many amazing cookbooks and online recipes that focus on mushrooms and may inspire you to experiment. As an introduction to what is possible, take a closer look at some of the mushrooms shown in Figure 12-3:

>> **Chanterelles** are sometimes described as having a fruity or peppery flavor. They pair well with meat and fish.

>> **Morels** have a meaty texture and a rich, earthy flavor. They are divine in a cream sauce over pasta.

- » **Oyster** mushrooms have a very mild flavor once they are cooked. Their chewy texture makes them a good replacement for meat in recipes.

- » **Enoki** mushrooms are delicate and cook quickly. You can cut off the base of the cluster and drop the fresh mushrooms into hot soup, and let the broth cook them.

- » **Shiitake** mushrooms are often available in dried form in grocery stores. To use them, pour boiling water over the dried mushrooms and allow them to rehydrate. Shiitake mushrooms make a nice addition to stir-fry and soups.

- » **Maitake** or hen-of-the-woods grows in large, intricate lumps. Their thin caps become deliciously crispy when these mushrooms are fried.

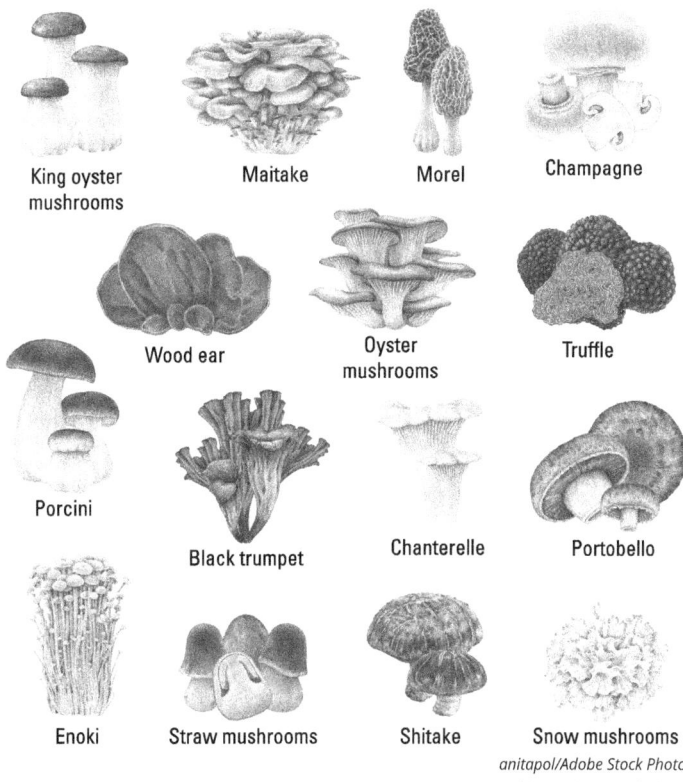

King oyster mushrooms   Maitake   Morel   Champagne

Wood ear   Oyster mushrooms   Truffle

Porcini   Black trumpet   Chanterelle   Portobello

Enoki   Straw mushrooms   Shitake   Snow mushrooms

FIGURE 12-3: A variety of edible mushrooms.

Although more varieties of mushrooms are making an appearance in grocery stores, they're often expensive. Two ways to reduce the cost are to go forage for them yourself (see the following section) or grow them yourself at home. (See Chapter 15 for details.)

Foraging to find wild fungi

People have many reasons for heading out into the wild to forage for fungi:

>> They might come from a culture that has a long tradition of foraging for fungi.

>> Some people, including gourmet chefs, want to experiment with new flavors that can only be found in wild-harvested foods.

>> Other people like the challenge and thrill of the hunt.

>> And finally, hunting fungi is a great way to get outside and connect with nature.

If you aren't someone who was brought up mushroom hunting, you may be uncertain about how to find the mushrooms you're interested in or how to make sure a mushroom is safe to eat. If you are planning on doing any foraging, it's important to make sure you have a good mushroom identification guide for your area. It's also a good idea to pair up with people who have experience. Many communities have mushroom clubs made up of people who love mushrooms. Members of these clubs sometimes lead groups to go mushroom hunting. (See Chapter 18 for information on the North American Mycological Association and how to locate your local mushroom club.)

To find a certain mushroom, get to know what it likes. Find out if it grows in your area and what type of environment it likes. Some mushrooms grow near certain tree species. Some like to grow in areas that have burned recently. Find out what time of year it usually appears. Familiarize yourself with its characteristics and whether it has any dangerous look-alikes.

WARNING

If you aren't sure about the identity of a mushroom, don't eat it!

Mushrooms are the visible reproductive structures of a much larger underground organism. In a sense, when you pick a mushroom, it's more like you are picking a fruit from a tree. Just like you don't hurt the tree if you pick all the fruit, you won't hurt the fungus if you pick all the mushrooms, especially if you leave a few of the older ones behind to spread the spores. So, in general, you can pick all the mushrooms you want from a population you find without hurting the fungus. You should know the rules of the area you're in, though, as some parks have limits on how much you can take.

One rule of ethical harvesting is always to take only what you will use.

TIP

In addition to this rule, here are a few things you can do to be kind to the mushrooms and other mushroom hunters:

>> **Leave some for other hunters.** You won't hurt the fungus if you take them all, but you might disappoint other mushroom hunters.

>> **Tread lightly among the mushrooms.** If you trample and compact the soil where fungi grow, you may reduce the future production of new mushrooms.

>> **Leave no trace.** This is just a basic rule of any activity that takes you into nature. If you pack it in, pack it out. And if you make a mess while cutting mushrooms, tidy up the area before you go.

In addition to your mushroom identification guide, there are some other useful tools to bring with you on the hunt. The first is a knife that you can use to cut away any dirt that's clinging to the base. The second is a bag or basket to hold your bounty. Many foragers like to use mesh bags like those used for laundry because they let air in, keep dirt, and let spores escape to start new mushrooms. Plastic bags aren't recommended because they keep in too much moisture, which can cause mushrooms to spoil more quickly. Speaking of dirt, you probably want to bring a brush with you so you can clean your mushrooms in the field. If you leave dirt on your mushrooms for too long, it can get stuck and be hard to remove. Some mushroom knives have a built-in brush, but you can also use a pastry brush, paint brush, or even a soft toothbrush.

Don't wash your mushrooms until right before you're going to cook them. Mushrooms absorb water, and washing them will make them waterlogged and more likely to spoil.

Once you get your mushrooms home, store them in your fridge loosely packed in a paper bag. If you've ever bought mushrooms in a shrink-wrapped package from the grocery store, you were being given a lesson in how not to store your mushrooms. Covering them in plastic holds in moisture and shortens their storage time. If you start to see dark, wet-looking spots on your mushrooms, they are starting to spoil and probably won't taste good anymore.

Keeping your mushrooms cool, dry, and airy will keep them fresh longer.

Fermenting with Fungi

Every culture around the world eats some type of fermented food, from bread to cheese to fermented beverages to various pickles and other vegetables. Modern food scientists in industrialized nations carefully control the fermentation of foods by sterilizing all the equipment, maintaining specific temperatures, and inoculating the food with specific microbes.

The first fermented foods, however, spontaneously occurred by the action of microbes naturally found on the surface of foods. Although some microbes will spoil food and make you sick, microbes that use fermentation can produce products that change food in ways that many people find delicious, like those shown in Figure 12-4. As a plus, the acids produced during fermentation keep food spoilage organisms from growing, preserving the fermented foods for relatively long periods even in the absence of refrigeration.

Fermentation is a metabolic process that allows cells to transfer energy from food molecules. Fermentation is an *anaerobic* process, which means cells don't need oxygen to do it (*an*=without, *aero*=air).

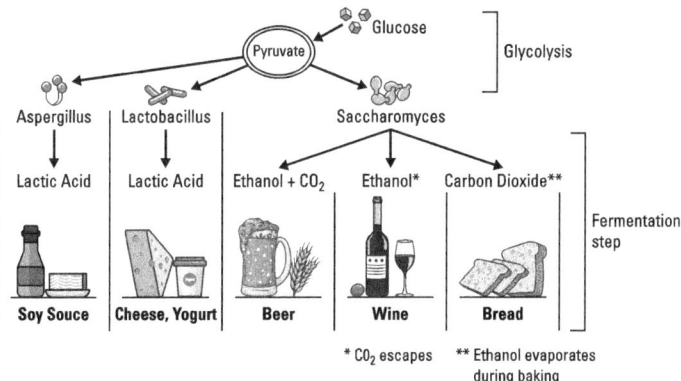

FIGURE 12-4: Examples of different types of fermentation.

Although some of the details differ, all types of fermentation involve at least two main steps:

>> **Glycolysis:** Cells break down the simple sugar glucose into a smaller molecule called pyruvate.

>> **Fermentation step:** During the fermentation step, cells convert pyruvate into various *fermentation products* such as lactic acid, ethanol (ethyl alcohol), and carbon dioxide (CO_2). Acidic fermentation products give fermented foods their tangy taste, CO_2 creates bubbles, and ethanol gives alcoholic beverages their mood-altering effects.

Microbes can undergo various forms of fermentation, some of which are illustrated in Figure 12-4. Baker's yeast and brewer's yeast are fungi in the genus *Saccharomyces* that do alcohol fermentation, producing ethanol (ethyl alcohol) and carbon dioxide as waste. Lactic acid bacteria, such as *Lactobacillus*, produce lactic acid through lactic acid fermentation, which is the same process used by your muscle cells when they are low on oxygen. The fungus *Aspergillus* can also do lactic acid fermentation.

Using yeast to make beer, wine, and bread

Yeast may be the most important microbe in human food production. As the essential organism for the production of beer, wine, and bread, it's definitely the most popular. Archeological evidence shows that people first started baking bread from grains about 14,000 years ago, but these breads were flatbreads that didn't contain any yeast. Nobody knows exactly how yeast got involved, but some people speculate that mixing flour with water to make flatbread led to fermented grain beverages (beer), and that a happy accident led to beer splashing back into a bread dough to create a dough that had enough yeast in it to cause the dough to rise. We'll never know if this is actually how it happened, but we do know from archeological evidence that leavened bread appeared about 9,000 years ago — 5,000 years after the appearance of flatbreads.

Yeast does ethanol fermentation, converting the sugars in food into ethanol (ethyl alcohol). Along the way, carbon atoms are removed from the sugar and released as part of carbon dioxide (CO_2). Carbon dioxide is a gas that causes the bread dough to rise, resulting in a light and airy loaf of bread.

Like all fermentations, ethanol fermentation consists of two main events, glycolysis and a fermentation step. The yeast uses glycolysis to break down the sugar glucose and extract the energy they need for growth, producing the molecule pyruvate as a result.

Two important events occur during glycolysis:

>> **Energy transfer:** Yeast cells transfer energy from glucose to a molecule called ATP (adenosine triphosphate). The yeast can use ATP to provide energy for growth, repair, and movement.

>> **Electron transfer:** Part of the energy extraction process involves transferring electrons from glucose to a molecule called NAD+ (nicotinamide adenine dinucleotide). When NAD+ accepts the electrons, its structure changes. To reflect this change, scientists call the new form NADH.

REMEMBER

The purpose of glycolysis is to obtain energy from glucose.

To keep growing, the yeast cells need to extract energy from more glucose molecules by repeating glycolysis. However, to perform glycolysis, the yeast must have more NAD+, not NADH.

The purpose of the fermentation step is to recycle NADH back to NAD+.

During ethanol fermentation, the yeast removes carbon atoms from pyruvate, releasing them as part of CO_2. Then they transfer electrons from NADH onto the remaining molecule, resulting in the production of ethanol.

Beer and wine officially appear in the archeological record soon after the appearance of leavened bread. Spontaneously fermented beverages go back into prehistory, but the oldest breweries appeared between 6000 and 3000 BCE (Before Common Era) in Turkey and Egypt. Grape cultivation and wine making appeared around the same time in the region around Mesopotamia. The ancient Egyptians recorded the process of winemaking on clay tablets that date back to 2031 BCE.

Yeast do the same fermentation process during beer and wine production. The differences in the beverages result from the starting ingredients — grain for beer and grapes for wine — and the production process. Beer is bottled or put into kegs before fermentation finishes, so some CO_2 remains trapped in the liquid. Wine, on the other hand, is usually bottled after fermentation is complete and all the CO_2 has been released into the air. For sparkling wines, additional sugar is added at the time of bottling to restart the fermentation process and create new bubbles.

SAVING THE REPUTATION OF FRENCH WINE

French wine is known today as some of the best wine in the world. Back in 1863, however, the reputation of French wine was on shaky ground. At the time, producers of fermented foods relied mostly on spontaneous fermentation, which resulted from the growth of microbes on the surface of the food. To make wine, for example, winemakers crushed grapes, then put the juice in vats to ferment without much control over the process. Sometimes, this resulted in wonderful wines, but other times the wine spoiled, turning into

(continued)

something sour or bitter. The inconsistency of French wine threatened both the wine industry and the country's economy. But then, Napoleon III called on Louis Pasteur, a French scientist who was well known for his studies on microorganisms. Pasteur had already done research into fermentation, showing that it was an anaerobic process. He'd also done some work on food spoilage and noticed that different microbes appeared under his microscope when he sampled a successful fermentation versus a spoiled food. Pasteur spent two years sampling wines and figuring out the causes of all the different "wine diseases" (as the French called them). He realized that if certain bacteria got into the grape juice, they would spoil the wine. He also knew that if he heated up the grape juice, he could kill the bacteria. Unfortunately, he might also ruin the flavor of the developing wine. So, he experimented with different heating protocols, raising the temperature on the grape juice just a little bit at a time, until he developed the perfect protocol of how long and how much to heat the grape juice to kill the wine-spoiling bacteria without ruining the wine. The result? Pasteurization was born, and French wine became celebrated throughout the world. We still use pasteurization today to control the microbes in many foods, from milk and juice to eggs, flour, and, of course, beer and wine.

Producing sake and soy sauce with aspergillus

About 5,000 years ago in China, people grew soybean crops for food and animal feed. Because soybeans spoil easily, they were placed in barrels with salt added as a preservative. Over time, the beans fermented much like pickles or sauerkraut. As they ferment, the soybeans turn into a paste called miso. Miso paste is easier to digest than the unfermented soybeans, and people have been eating it for centuries.

About 500 years ago, someone discovered that instead of discarding the salty liquid at the bottom of the soybean barrels, they could use it for cooking. Thus, soy sauce was invented. Soon, soy sauce was shipped in barrels throughout Asia and made its way to Europe in bottles by the 1600s. Today, soy sauce is used all over the world.

Soy sauce is brewed in two stages:

1. First, the soybeans are steamed and mixed with toasted, crushed wheat. Two molds in the genus *Aspergillus*, *A. oryzae* and *A. sojae,* are added to the mixture to make *koji*. The koji is left uncovered for a couple of days. Once *Aspergillus* has broken down the large molecules in the soybeans and wheat into small molecules, the koji phase ends.

2. Next, salt and water are added to the koji to form a mash called *moromi*. Moromi is then put in airtight containers where it is allowed to ferment for at least six months. The mash is squeezed to get the liquid soy sauce. Finally, the sauce is filtered, pasteurized, and tightly bottled for distribution.

Using molds to preserve milk as cheese

There's a saying that "cheese is preserved milk." This saying makes sense if you compare how quickly milk versus cheese would spoil if you left them out on the counter. Like the other fermented foods presented in this section, cheese was probably discovered spontaneously when someone left milk out. Microbes fermented the milk, and a person decided to do what they could with the result.

The process of converting milk into cheese occurs in several basic steps:

1. Bacteria ferment the sugars in milk using lactic acid fermentation(refer to Figure 12-4), which produces lactic acid as the fermentation product.

2. The acid causes the proteins in the milk to coagulate, forming solid lumps called *curds*.

3. The cheesemaker separates the solid curds from the liquid whey.

4. The cheesemaker presses the solid curds into a cheese mold to shape the cheese.

5. The cheesemaker salts the cheese with a brine solution and puts it in a cool, dry place to age.

Many factors contribute to the differences between one cheese and another. The type of milk, the speed of coagulation, and how much whey is pressed out of the curd all contribute to the final form of the cheese. For faster coagulation, cheesemakers may add *rennet*, a substance that contains an enzyme that speeds up the process.

Aging also affects the flavor of cheese. As cheeses age, bacteria continue to break down the proteins, which alters the flavor and texture of the cheese. Molds may also contribute to the aging and final flavor of some cheeses:

>> **Blue cheeses:** Mold spores are added to the curd before it begins to age. The cheesemaker pokes holes into the curd to allow the mold to grow through the cheese and develop the classic blue veins. Roquefort cheese contains the mold *Penicillium roquefortis,* which is only found in the caves in the Roquefort region of France.

>> **Bloomy rind cheeses:** Cheeses like brie and camembert are sprayed with mold before they begin to age. The mold grows, first forming a soft rind on the outside of the cheese, then working its way inward, digesting and changing the bland curds into soft, creamy, flavorful cheese. The mold used to make brie is often *Penicillium candidum*, while camembert is made with *Penicillium camemberti*.

TAKING CHEESE SERIOUSLY

The French are serious about their cheese. Like many French foods, different types of cheese are tied to the region in which they are made. This concept, called *terroir,* represents the idea that a particular food is influenced by its environment. In the case of cheese, for example, the flavor could be influenced by the type of cow that produces the milk, the food the cows eat, and the weather during the grazing season. So, each region in France has its traditional cheeses that can only be produced in that region.

Along with terroir, each type of cheese has a specific process for its manufacture and rules that go along with that process. These rules are based on the traditional methods for making each cheese that have been passed down through many generations. The rules include things like the type of cow and their foraging routine, including how much grazing area per cow is necessary. In some cases, the cheese production is split into three defined roles: the dairy farmer who raises and milks the cows, the cheesemaker who turns the milk into cheese, and the cheese ager who monitors the cheese during the aging process. Once the cows are milked, there's a time limit for turning the milk into cheese, often as little as 24 hours. A cheesemaker who produces Camembert said he has to follow 63 pages of rules! For people who love cheese, though, the effort is worth it. The French produce some of the best and most diverse cheeses in the world.

Chapter **13**
Spiritual Uses of Fungi

As a consequence of gathering and eating mushrooms, the earliest people probably experienced unexpected outcomes, including illness, death, and mind-altering experiences. Archaeological evidence suggests that these people saw mushrooms as important and powerful, and some anthropologists think hallucinogenic mushrooms may have shaped some of our myths. In this chapter, you look at how your brain makes sense of the world and how some chemicals from mushrooms can change your perceptions.

Exploring Human Perceptions

It can be easy to oversimplify how you perceive the world around you: You see with your eyes, hear with your ears, smell with your nose, taste with your tongue, and touch with your fingers, right? Well, not quite. It's true that your *sense organs* — your eyes, ears, nose, tongue, and skin — contain *receptors* that detect signals from your environment, but how you make sense of those signals is all about your brain.

Sensation is the intake of information by your sensory receptors. *Perception* is how your brain organizes and interprets that information.

Optical illusions illustrate the difference between sensation and perception. The illusion in Figure 13-1, for example, works because your brain can focus on certain information as it tries to make sense of what you see. The image doesn't change, but how you interpret it does. Optical illusions like this one tap into different aspects of how our brain processes information, tricking our brain into thinking it sees something that isn't there.

Peter Hermes Furian/Adobe Stock Photo

FIGURE 13-1: The rabbit-duck illusion.

Scientists have been studying the human brain for hundreds of years in an attempt to identify how it functions. Based on the physical structure of the brain, early scientists made a map that divided the surface of the brain into regions called *lobes,* like those shown in Figure 13-2. Doctors treating people with injuries to specific areas of the brain noticed that damage to certain areas seemed to correspond to difficulties in particular tasks, such as walking or speech. More recently, scientists have measured how electrical signals and blood flow in the brain change in response to specific stimuli. Although all parts of your brain work together, certain parts of the brain seem to specialize in particular functions.

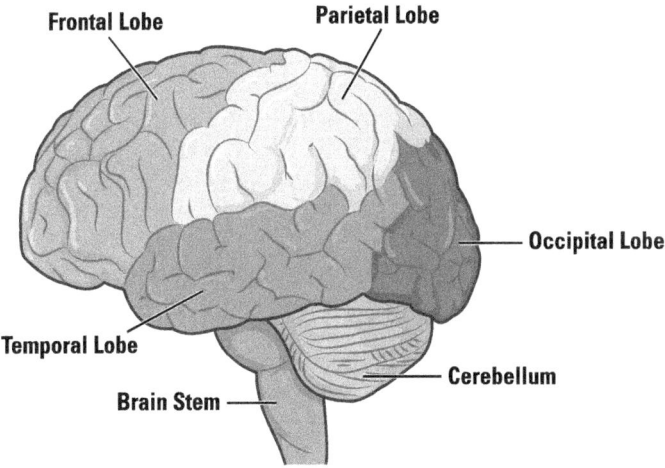

FIGURE 13-2: Some regions of the brain and their primary functions.

Sending and receiving signals with neurons

The two most common types of cells found in your brain are nerve cells, or neurons, and glial cells. Neurons send and receive signals, while glial cells support the function and health of neurons.

REMEMBER

A *neuron*, or *nerve cell*, is a specialized cell that sends information from the body to the brain and then from the brain back to the body.

Neurons have a distinctive shape that's important to their function. They're very long, with a rounded *cell body* that contains the nucleus and a long, tail-like *axon* that extends away from the cell body. When a neuron receives a signal, it reacts by sending a weak electrical pulse down its axon. When the signal reaches the *axon terminal*, the neuron can pass the signal to another cell. Scientists think the length of neurons allows them to process information while passing on the signal.

The complex network of nerves in your body forms your *nervous system*, which can be divided into two main parts:

>> Your central nervous system is your brain and spinal cord.

>> Your peripheral nervous system is the nerves that move from your spinal cord to your body, plus other nervous tissue in the body.

Your nervous system regulates the functions of your mind and body. It takes signals in from your sensory organs, processes the signals, and sends response signals to your muscles and organs. Without you even noticing, it maintains essential functions such as your breathing, heart rate, digestion, and organ function. You also need your nervous system to move your muscles, think, and learn. Your nervous system can even trigger your emotions.

Moving across synapses with neurotransmitters

Groups of neurons use a combination of chemical and electrical processes to send signals along pathways in your nervous system. Neurons make chemicals called neurotransmitters in their cell bodies, then ship them down to their axon terminals in transport vesicles. (For a refresher on vesicles, head back to Chapter 4.)

REMEMBER

Neurotransmitters are molecules produced by neurons to send signals to other cells, such as neurons, muscle cells, or gland cells.

To send a signal, the vesicles of neurotransmitters fuse with the cell membrane in the axon terminals, releasing the neurotransmitters into a space called a synapse.

REMEMBER

A *synapse* is a very small gap between a neuron and another cell.

Messages pass through your nervous system at incredible speeds. One neuron can receive a signal and pass it to another neuron in a series of steps that only take about two milliseconds (0.002 seconds) to complete:

1. The *presynaptic neuron* receives a signal, causing it to send a weak electrical signal down its long *axon*.

2. When the electrical signal reaches the *axon terminals* at the end of the cell, the presynaptic neuron releases neurotransmitters into the synapse.

3. The neurotransmitters travel across the synapse, then bind to receptors on the surface of the *postsynaptic neuron*.

4. When the neurotransmitter binds to its receptor on the postsynaptic neuron, it triggers an electrical signal through the neuron, and the process repeats itself.

The relationship between neurotransmitters and their receptors is specific. Each neurotransmitter fits into certain receptors like a key fits into a lock.

Each neurotransmitter has a specific effect on the target cell. They may excite the target cell, causing it to pass on the signal, or they may inhibit it, stopping the message from traveling further along the pathway. Some neurotransmitters can also affect how the target cell communicates. Depending on their action and their target cells, each neurotransmitter affects certain functions in the body. Here are a few of the major players in your nervous system:

>> **Acetylcholine** excites neurons, influencing muscle contraction, heartbeat, and sweating.

>> **Glutamic acid** excites neurons, playing an important role in cognitive functions like thinking, learning, and memory.

>> **Gamma-aminobutyric acid (GABA)** inhibits neurons, regulating your mood with a calming effect. It's important for the regulation of anxiety and depression.

>> **Serotonin** inhibits neurons and affects your mood, behavior, sleep, and memory. It helps control anxiety, sexual arousal, and appetite.

>> **Dopamine** can excite or inhibit neurons and plays a role in memory, learning, and behavior. It's connected with feelings of pleasure and reward.

Now that you've explored some of the fundamentals of how your nervous system regulates your body and mind, head to the next section to learn how people use mushrooms to alter their perceptions.

Expanding the Mind with Magic Mushrooms

Archaeological evidence suggests that people have a long history of collecting and using hallucinogenic mushrooms. Ancient cave paintings found in Spain and Algeria that are six to eight thousand years old show people holding mushroom-shaped structures that some people think represent the magic mushrooms in the genus *Psilocybe*. Rock paintings from Siberia show people that appear to be dancing with mushrooms on their heads.

In addition to this ancient evidence, more recent studies have revealed some of the ways indigenous people around the world use hallucinogenic mushrooms for spiritual purposes. These traditions have been handed down through many generations and support the idea that hallucinogenic mushrooms have been important to human societies for thousands of years.

REMEMBER

A *hallucinogen* is a substance that alters a person's perception of reality. If a hallucinogen is taken for religious or spiritual purposes, it's called an *entheogen*.

Unlocking the power of psilocybin

Psilocybe is a genus of small brown mushrooms in the phylum Basidiomycota (see Chapter 11). They grow all around the world, with different species found in each region. Although the appearance of each species varies slightly (you can see one species in Figure 13-3), they have a few very recognizable characteristics:

>> **Blue bruising:** If you press firmly on the cap of most *Psilocybe* mushrooms, a blue color should appear. (A few non-hallucinogenic species of *Psilocybe* don't react with blue bruising.)

>> **Purple-brown spores:** If you make a spore print on a light colored surface, the spores should have a purple-brown color (see Chapter 4 for instructions).

anitapol/Adobe Stock Photo
FIGURE 13-3: *Psilocybe cubensis.*

Most *Psilocybe* mushrooms produce a chemical called *psilocybin* that has hallucinogenic effects on people. The mushrooms became well-known in the English-speaking world during the 1960s counter-culture movement, and people started calling them "magic mushrooms."

When people eat magic mushrooms, our bodies change the chemical psilocybin slightly, producing the chemical psilocin (you can compare the two molecules in Figure 13-4).

Our cells convert psilocybin into psilocin by removing a phosphate group.

Psilocin is the active chemical in the human brain, where it behaves like a neurotransmitter.

Psilocin is a serotonin *agonist,* in other words, a molecule that behaves like serotonin.

If you compare the chemical structures of psilocin and serotonin in Figure 13-4, you can see they have a similar structure. Remember that serotonin is a neurotransmitter that sends signals by binding to receptors on cells, fitting into those receptors like a key into a lock. The structure of psilocin is so similar to serotonin that it can fit into the same locks. So, if a person eats psilocybin, the result is a rush of a serotonin-like molecule into the brain.

Hallucinogens that alter a person's perception of awareness by mimicking increased serotonin are called *psychedelics (psyche*=soul, *delein*=reveal*)*.

Psilocybin   **Psilocin**   **Serotonin**

FIGURE 13-4: Psilocybin, psilocin, and serotonin.

TIP

You're probably familiar with the word *antagonist,* which means something that opposes. Well, if you take the "anti-" (against) out of antagonist, you're left with "agonist." So you can remember the term agonist as something that doesn't oppose (officially, it's something that binds to a receptor and triggers the same reaction as the molecule that usually binds to the receptor).

In high doses, psilocybin completely changes a person's perception of reality. People who have taken psilocybin often say that it made them introspective and gave them new perspectives on their lives. Although scientists are still studying how psilocybin causes these changes in perception, they have started to put together some pieces of the puzzle:

>> One study showed that psilocybin stimulated activity in the amygdala of the brain. The amygdala is a pair of almond-shaped structures located in the temporal lobes of the brain. It processes emotions, especially fear and anxiety, and is involved in forming memories.

>> Psilocybin changed the pattern of *functional connectivity* between areas of the brain. Functional connectivity occurs when two areas of the brain that are separate from each other show the same pattern of activity at the same time, as if they were connected. The greatest changes occurred in the areas of the brain known as the *default mode network,* which are usually active when a person is resting or not focused on an outside stimulus. Scientists think that the default mode network is important in creating a person's sense of space, time, and self.

FLESH OF THE GODS

When the Spanish colonizers came to Central America in the 1600s, they encountered indigenous people like the Aztecs and the Mayans, whose cultural practices were very different from those of Europe. The Catholic priests were horrified by religious practices that included the consumption of mushrooms to induce an altered state of consciousness. The Aztecs called the mushrooms *teonanactl,* which means "flesh of the gods" in the Nahuatl language. The mushrooms, which were most probably *Psilocybe mexicana, P. cubensis,* or other *Psilocybe* species, induced hallucinations that were part of the spiritual

ritual. The Spanish people who observed these rituals reported that the mushrooms were sometimes mixed with honey, mescal, or chocolate, and that music and dancing were sometimes part of the ceremony. Spanish observers said that the people consumed the mushrooms to have visions or speak with their God.

The Spanish persecuted the people who practiced these mushroom-based religions, so their voices are lost to us. We are left to piece together their story from archeological evidence, writings from Europeans, and the practices of their descendants (see the section "Practicing traditional healing" for details). We know that mushrooms and other entheogens were important to them. The Aztec god of flowers, Xochipilli, is decorated with entheogenic plants, and he has mushroom caps on his earlobes and knees. The Mayans carved "mushroom stones" that look like people or animals with mushroom caps on their heads. Images of mushrooms have also been found on Mayan artifacts associated with religious practices.

Practicing traditional healing

Although there is evidence that indigenous groups around the world used hallucinogenic mushrooms in spiritual rituals, colonization, climate change, and industrialization disrupted many traditional practices, leading to the loss of much of this history. Modern accounts of the spiritual use of *Psilocybe* are sparse but have been documented in some areas. The Basotho people of Lesotho include *Psilocybe* in a hallucinogenic drink that they use for spiritual healing. *Psilocybe* is also used in healing rituals in the Oaxaca region of Mexico.

In the 1950s, a wise woman from Oaxaca named Maria Sabina agreed to perform a spiritual ritual for an American business-man and mushroom enthusiast named Robert Gordon Wasson. After Wasson returned to the United States, he wrote an article about his experience for *Life Magazine*. From his account and the writings of others who knew Maria Sabina, we have some insight into the importance of *Psilocybe* to spiritual rituals at that time. Unfortunately, Wasson's article also sparked a flood of tourism to Maria's remote village from people seeking the mind-altering effects of the "magic mushrooms." Maria suffered in her community as a result and was even temporarily imprisoned when the Mexican government criminalized the use of *Psilocybe*.

Maria Sabina was a Mazatec woman who grew up in the Sierra Mazateca mountains in Oaxaca, Mexico. Her father's family practiced traditional healing, and Maria witnessed rituals involving mushrooms when she was growing up. When she was an adult, her sister became ill, and the local healers had no hope. Maria turned to Los Niños Santos, or the Holy Children, which is what she called the mushrooms. She used the mushrooms to conduct a healing ritual, and her sister recovered. After that time, Maria became known as a *sabia*, or wise woman.

After Wasson was introduced to Maria Sabina, he and a friend were allowed to participate in a spiritual ceremony called a *velada*. In the days that followed, they wrote down everything they could remember about the ceremony and their experiences. Here are some of their observations on the ritual itself:

» Participants were served chocolate before the ceremony, which took place at night.

» After Maria cleaned the mushrooms, she passed them through the smoke of some burning incense and said prayers over them.

» Maria sat on a mat before an altar with a picture of the baby Jesus and other Christian images.

» She and her daughter both ate mushrooms along with most of the participants.

» She used a flower to snuff out the only candle in the room. After that, all participants remained in the dark until morning.

» Maria chanted throughout the night, calling on the mushrooms in the name of Jesus Christ and the saints to speak. She also asked for healing for her sick son.

Maria spoke her indigenous language of Mazatec, so her words had to be translated for the English speakers. Other Spanish-speaking writers talked with Maria later in her life and wrote about her beliefs and practices. One of them said that Maria's power came from her chants, which were a kind of poetry. Some of her chants have been published in Spanish and English translations.

Managing mental health

The healing properties of magic mushrooms may include their ability to improve mental health. During the 1960s, there was an initial burst of research on the use of psychedelics to treat patients with mental health challenges. But when people started using these drugs recreationally, governments reacted by banning their use. In the United States, for example, *Psilocybe* was listed as a Schedule 1 controlled substance.

REMEMBER

Schedule 1 controlled substances are defined as having no medical use and a high risk for abuse.

Once *Psilocybe* and other psychedelic drugs like lysergic acid diethylamide (LSD) were classified as Schedule 1 substances, research into their medical use ground to a halt.

Recent policy changes have made it easier for researchers to use psychedelics in clinical trials for mental health. In many of these studies, participants take a large dose of psilocybin in the presence of a psychiatrist. The patients remain with their psychiatrist throughout the experience and discuss it extensively afterward. Although this research is in early stages, psilocybin shows promise for the treatment of depression, obsessive-compulsive disorder, and alcoholism. One drawback is the potential for side effects such as headache, nausea, and anxiety.

Another potential option for using psilocybin to treat mental health is to use small doses instead of large ones. This technique, called *microdosing*, aims to give a patient enough of the drug to give a benefit but not enough for them to feel an effect. Reviews of the impact of microdosing suggest that it may be beneficial in controlling anxiety, boosting creativity, and improving mood, but there are only a few controlled scientific studies. One study on microdosing with *Psilocybe* mushrooms didn't show significant differences in physical activity or creativity between people who took the mushrooms and people who didn't, but more research in this area is needed.

Avoiding dangerous impostors

One of the advantages of *Psilocybe* mushrooms is that they have a low toxicity. They may alter your perception of reality, and may even have unpleasant side effects, but they are unlikely to kill you unless you ingest very large amounts or take them in combination with other drugs.

WARNING

In the United States, it's illegal to possess psilocybin and psilocin under federal law. This means that it can also be considered illegal to possess species of *Psilocybe* that contain these chemicals, and most state courts in the United States have supported that interpretation of the law. Some states allow the possession of *Psilocybe* spores because they don't contain psilocybin or psilocin, but other states have passed laws to specifically prohibit spore possession. Other states have either recently passed or are considering laws designed to allow the use of *Psilocybe* in therapeutic settings. If you are interested in *Psilocybe* for any reason, be sure you investigate the laws for your area.

WARNING

If you ever try to harvest *Psilocybe* from the wild, be on the lookout for a dangerous impostor. The mushroom *Galerina marginata* is also a small brown mushroom that superficially resembles *Psilocybe*, but it can be deadly (compare Figure 13-3 and the *Galerina marginata* shown in Figure 13-5). *Galerina marginata* can even be found growing among magic mushrooms, so you need to be very careful about what you are harvesting.

Galerina marginata has rusty brown spores, while *Psilocybe* species have purple-brown spores. Also, hallucinogenic species of *Psilocybe* will show the blue bruising reaction.

Another potential danger comes from thinking that all psychedelic mushrooms are the same. Many people have tried taking the fly agaric *Amanita muscaria* (shown in Figure 13-5) because they hear it's psychedelic.

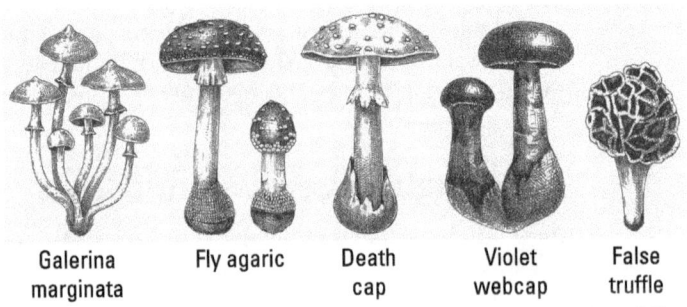

| Galerina marginata | Fly agaric | Death cap | Violet webcap | False truffle |

sketched-graphics/Adobe Stock Photo

FIGURE 13-5: Poisonous mushrooms.

REMEMBER

Amanita muscaria is not the magic mushroom. It is psychedelic, but it and other *Amanita* species are responsible for most cases of mushroom poisoning.

Fortunately, *A. muscaria* is usually recognizable because of its orange-red cap with white warts. Sometimes, however, rain or washing will remove the warts, and A. muscaria can be mistaken for the edible species *A. caesarea*. You can double-check your identification by looking for a volva — *A. caesarea* has a volva, while *A. muscaria* doesn't. Ingesting *A. muscaria* can result in disorientation, agitation, changes to blood pressure and heartbeat, and loss of consciousness. Severe poisoning cases may result in coma or even death.

A. muscaria and other psychedelic species of *Amanita* have different effects on the body because they contain different chemicals than *Psilocybe* mushrooms. The hallucinogenic chemicals in *Amanita* species are muscimol and ibotenic acid. When *Amanita* species are ingested, the body converts ibotenic acid into additional muscimol. These chemicals move into the brain where they behave as neurotransmitter agonists.

TECHNICAL STUFF

Ibotenic acid is a glutamic acid agonist that binds to receptors for glutamic acid. Muscimol is a GABA agonist that binds to GABA receptors.

After eating *A. muscaria*, people feel the effects of ibotenic acid first, such as hallucinations, confusion, and agitation. Muscimol also triggers hallucinations and begins to depress the nervous system, making people feel lethargic. In severe poisoning cases, the muscimol can trigger coma.

WARNING

A recent topic circulating on the Internet is the idea *A. muscaria* is an "edible" mushroom, meaning it is safe to eat and not poisonous. These posts point out that the toxins are water-soluble and then give directions for boiling the mushrooms to make them safe to eat. It's true that the toxins are water-soluble and that it's possible, if you are very careful and precise, to make *A. muscaria* safe to eat. You need to be careful about the volume of water versus the weight of mushrooms, plus you need to remember to change the water and boil the mushrooms twice. If the mushrooms require this much work or you will suffer the consequences, then are these mushrooms truly edible? Or is it, as they say, "You can eat anything. Once."

MAGICAL BREWS AND LEGENDS

Human history is scattered with interesting folk tales and legends that suggest the involvement of psychedelic mushrooms. In Ancient Greece, pilgrims journeyed from Athens to Eleusis to participate in the Eleusinian Mysteries, a ritual in honor of the goddess Demeter and her daughter Persephone, who had to travel to the underworld for part of the year to be with her husband Hades. The pilgrims would fast and then drink a brew called kykeon, which triggered visions and ecstasy. Although no one knows for sure what exactly was in kykeon, the main ingredient was probably barley flour. Some scientists think the barley used to make the flour was infected with the ergot fungus, *Claviceps purpurea*. This idea is supported by some evidence, including a picture of Demeter holding an ergot-infected sheaf of grain on a piece of Greek pottery. *C. purpurea* contains several hallucinogenic chemicals, the most famous of which is lysergic acid diethylamide (LSD).

Another more recent proposal is the idea that Santa Claus may have been inspired by *Amanita muscaria*. Red and white? Check. Flushed and rosy? *Amanita* can do that. Flying reindeer? Shamans in Siberia sometimes include *A. muscaria* in their toolkit of trance-inducing substances, and somewhere along the way, someone suggested that this may also be true of the Sami people, who are nomadic reindeer herders. A quick search on the Internet will pull up reports on this idea from many news outlets, so it must be true, right? Well, not so fast. Scholars who've looked into this idea have pointed out that the Sami don't use *A. muscaria*. And even those Siberian shamans don't rely on this mushroom. It's just one thing they sometimes use. It seems much more likely that Santa Claus was based on the Dutch Sinterklaas, who watches kids to see if they're naughty or nice and then rides his white horse over the rooftops to leave presents in the good children's shoes.

Chapter **14**

Using Fungi as Medicine

P eople have been using fungi as medicine for thousands of years. In traditional Chinese medicine, mushrooms are recognized for their power to stimulate the immune system, support organ function, and promote longevity. Fungi produce powerful chemicals that are key to their healing ability. Scientists isolated some of these chemicals to produce the first antibiotics. Guided by traditional knowledge, research on fungal chemicals continues as scientists seek to develop new treatments for diseases such as cancer, diabetes, heart disease, and dementia. In this chapter, you find out about the healing properties of fungi, from the first antibiotics to the current development of fungal-based medicine.

Producing Antibiotics from Fungi

The first antibiotic developed for large-scale use in human populations was penicillin, which is produced by molds in the genus *Penicillium* (see Figure 14-1).

REMEMBER

An antibiotic is a chemical produced by a microbe that targets bacterial cells.

The discovery of penicillin and its development into a therapeutic treatment changed medical practice and the world. Doctors used penicillin to prevent the deaths of thousands of Allied soldiers from battlefield infections during World War II. Penicillin also contributed to post-war recovery by curing sexually transmitted infections like syphilis and gonorrhea. After the war years, penicillin was used to treat the general population for diseases including meningitis, pneumonia, strep throat, and scarlet fever.

The impact of penicillin goes beyond the diseases it was able to cure. It was the first antibiotic to be discovered, but it wasn't the last. The discovery of penicillin kicked off a rush of exploration for more antibiotics. In the 1940s, Selman Waksman and his students at Rutgers University screened thousands of strains of soil bacteria, ultimately isolating more than 15 different antibiotics. The screening protocols they developed are still the basis for how scientists screen for antibiotics today. Likewise, the culturing and purification methods developed to manufacture penicillin provided the foundation for future antibiotic production. At the end of the 19th century, approximately 33 percent of deaths were due to infectious disease. By the end of the 20th century, that number dropped to about 4 percent.

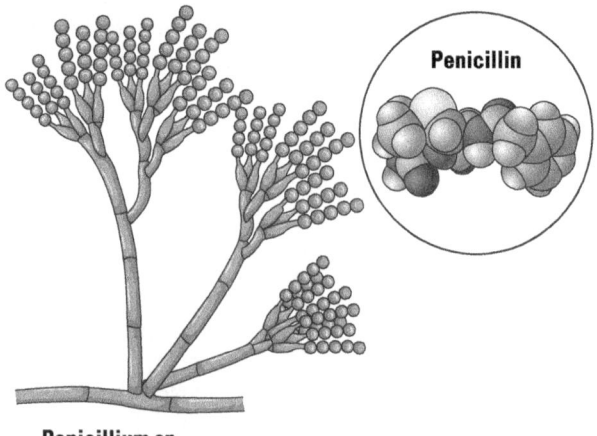

Penicillium sp.

FIGURE 14-1: *Penicillium* and penicillin.

Engaging in microbial warfare

Fungi live in complex microbial communities alongside bacteria and eukaryotic microorganisms. Many different relationships form between microbes in these environments, ranging from beneficial partnerships to competition for resources to outright predation. Sometimes it's literally a microbe-eat-microbe world.

In an environment like this, it's a definite advantage to have some tools that can give you a competitive edge or defend yourself against a neighbor who decides you might make a good meal. Microbes produce a wide variety of secondary metabolites that are useful in situations like these.

REMEMBER

Secondary metabolites are metabolic products that are produced in certain circumstances, whereas *primary metabolites* are metabolic products that are essential to basic growth and energy production.

For a fungus, primary metabolites would be things like cell wall components and enzymes needed to digest food. Secondary metabolites include pigments produced to protect the fungus from UV radiation and antibiotics to kill off the competition for food. To induce the production of secondary metabolites in the laboratory, scientists must figure out the right growth conditions. For example, many fungi produce secondary metabolites in response to stress, such as low food availability.

Identifying useful compounds

The power of penicillin was discovered by a happy accident, as described in the sidebar "Finding the first antibiotic," when Sir Alexander Fleming noticed that a clear barrier had formed between the bacteria he was growing in a culture dish and a contaminating mold that fell into the dish. He figured out that the mold was releasing a chemical that killed the bacteria. That chemical was penicillin and the bacterium was *Staphylococcus aureus*.

The observation of a clear zone between two microbes growing in culture is still used by scientists to look for useful antibiotics today. Once a potential chemical has been identified, scientists place a test dose of that chemical onto a paper disk and then place the disk onto a culture dish of bacteria. The chemical will spread, or diffuse, out of the disk onto the bacteria. If the chemical can kill them or stop them from growing, then a clear zone will appear around the disk as shown in Figure 14-2.

The *disk diffusion method* tests the ability of chemicals to stop microbial growth. Scientists measure the diameter of the clear zone, called the *zone of inhibition*, around the paper disk. The wider the zone, the more effective the chemical is at stopping microbial growth.

In Figure 14-2, the chemical in the disk with a 16mm zone of inhibition is more effective at stopping the microbe than is the chemical with a 12mm zone of inhibition. The chemical with no zone of inhibition doesn't inhibit the microbe at all.

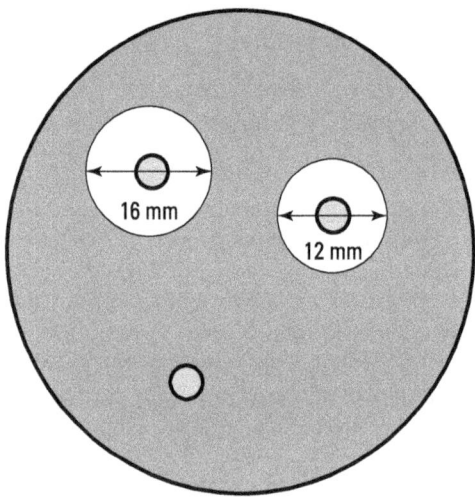

FIGURE 14-2: Zone of inhibition.

FINDING THE FIRST ANTIBIOTIC

In 1928, Professor Alexander Fleming returned from vacation and started cleaning up his lab. As he was looking through dishes in which he'd been growing the bacterium *Staphylococcus aureus*, he noticed something unusual. It looked like a fuzzy green mold had contaminated some of the cultures. That wasn't so unusual, especially because he'd been away. What he particularly noticed, though, was the clear zone that existed between the mold and the antibiotic. When several microbes are growing in the same dish, they usually grow right next to each other, or even on top of each other. But in this case, there was a very clear boundary in which no *S. aureus* was growing. Dr. Fleming grew cultures of the mold and determined that something it was producing, which he initially called "mold juice," could kill *S. aureus*. He tested the mold juice, which he named penicillin after the mold *Penicillium*, and found it could also kill other bacteria, including the ones that caused strep throat, meningitis, and diphtheria. He was unable to create a stable, purified solution of penicillin, but he wrote about his research in a paper published in 1929.

A lot of work remained to be done to move from Dr. Fleming's initial discoveries to the use of penicillin as a therapeutic agent. The next two scientists to step up to the plate were Dr. Howard Florey and Dr. Ernst Chain, who was employed as a biochemist in Dr. Florey's lab at Oxford University. Florey was interested in the interactions between bacteria and mold, and in 1938, he read a copy of Fleming's paper. Working together, Florey and Chain were able to grow larger quantities of the *Penicillium* mold and produce extracts that saved the lives of mice infected with *Streptococcus*. The scientists also tried to save the life of a police constable with a life-threatening infection. Their extract worked, but they just didn't have enough of it to kill all the bacteria, and the constable eventually died. It was clear that penicillin had potential, but they needed to come up with a way to increase production.

In 1941, Dr. Florey flew to Peoria, Illinois, with another biochemist from his lab, Dr. Norman Heatley. Their mission was to develop large-scale production of penicillin. One problem was that the original mold isolated by Fleming, called *Penicillium notatum*, didn't produce a lot of the life-saving chemical. The scientists got a lucky break when their lab assistant, Mary Hunt, came into work with a cantaloupe that was growing a different strain of *Penicillium*, called *P. chrysogenum*. When the scientists tested this mold, they found it made 200 times more penicillin than the original mold. Production was scaled up into huge industrial fermenters, making penicillin available to treat wounded soldiers during World War II. Drs. Fleming, Florey, and Chain shared the Nobel Prize in Physiology or Medicine in 1945 for "for the discovery of penicillin and its curative effect in various infectious diseases."

Continuing the search

The hunt for new antibiotics slowed down after the initial rush in the mid-20th century. During that time, which is sometimes called the "golden age of antibiotics," scientists discovered about

27 different classes of antibiotics, 5 of which are produced by fungi. Since that time, only eight new classes have been approved.

Several factors combined to inhibit the discovery of new types of antibiotics:

>> One reason for the slowdown was simply the success of antibiotics. Thanks to the explorers of the golden age, doctors had a large arsenal of antibiotics that very effectively treated infectious diseases. Some people didn't see the need for more.

>> As scientists continued to search using the same methods, they turned up fewer new types of antibiotics. Instead, they often recovered repeats of previously identified drugs. The time and expense remained the same, but the reward declined.

>> Pharmaceutical companies dedicated their resources toward more profitable drugs. Although governments fund some public health research, private companies also fund scientific inquiry. Increasingly, pharmaceutical companies in the United States allocate their research dollars to medications for chronic conditions because these tend to yield the highest profit margins.

Unfortunately, bacteria didn't get the memo that it was time to take a break. In fact, the use of antibiotics sped up the evolution of bacteria, leading to greater numbers of antibiotic-resistant strains. The change in bacterial populations was due to *evolution by natural selection*, the theory first proposed by Charles Darwin in 1859:

>> **Individuals in most natural populations produce more offspring than can survive.** Bacteria will keep multiplying as long as food is available, and then begin to die as food decreases and waste increases.

>> **Individuals have unique characteristics.** In any population of bacteria, genetic differences make the individuals a little bit different from each other. Some individuals may be more susceptible to an antibiotic than others.

>> **Some characteristics are inherited from parents to offspring.** Just like people, bacteria copy their DNA whenever they reproduce, passing copies of their DNA to their offspring.

>> **Organisms with characteristics best suited to their environment are more likely to survive and reproduce.** Whenever a population of bacteria encounters an antibiotic, the antibiotic kills the individuals in the population who are most susceptible first. With longer exposure, the more resistant bacteria may also be killed. However, if the antibiotic doesn't kill all the bacteria, then the most resistant bacteria will survive and reproduce, creating a new population that is more resistant than the original.

Antibiotics act as a *selection pressure* on populations of bacteria. In other words, they influence who lives and who dies, pushing the population to evolve in favor of antibiotic resistance.

The evolution of populations of antibiotic-resistant bacteria is one of the most challenging issues in medicine today. Penicillin, the original wonder drug, can no longer kill most strains of *Staphylococcus aureus*. Some strains of bacteria exist today that can no longer be killed by any available antibiotics. A study in the British medical journal *The Lancet* predicts that deaths from antibiotic-resistant infections are going to start rising sharply, increasing by 70 percent by 2050.

Scientists and medical professionals recommend a variety of tactics to avoid this outcome, including:

>> **Careful use of antibiotics.** This means only using antibiotics when they're necessary, using the correct antibiotics for a particular infection, and making sure the full dose is taken. For example, antibiotics don't work on viruses, so using antibiotics when someone has a viral infection may increase the speed of bacterial evolution without any benefit. Also, making sure a patient takes the full dose of antibiotics makes it more likely that all the infecting bacteria will be killed, rather than leaving room for the most resistant to survive.

>> **Widening the search pool for new antibiotic-producing organisms.** Scientists have two main strategies for this:

  • **Use genetic engineering to tap into the potential of organisms that have never been grown in the lab.** Scientists estimate that they've only identified 0.001 percent of the microbes on Earth, and they've grown fewer than that in the lab. With modern technology, however, scientists can extract DNA from environmental samples, chop it up into small pieces, and insert those pieces into microbes that do grow in the lab. Then, they can use traditional methods to screen these genetically engineered microbes for new antibiotics. All of the original discovery of antibiotics was done on organisms that could be grown in the lab, so this technique opens up the potential for finding new drugs from previously unknown organisms.

- **Search environments that have never been searched before.** Many of our antibiotics come from bacteria that grow in the soil. Scientists are still looking at soil samples, but now they're looking in unusual soil, like around Chernobyl, the site of the 1986 nuclear power plant disaster. The exposure to high levels of radiation triggered higher rates of mutation, or change, in the DNA of microbes in the area. Scientists are also looking in ocean water from the twilight zone, the layer of the ocean between 200 and 1,000 meters deep. Living things in this water have traits that allow them to survive high pressure, cold temperatures, and darkness, so they are very different from surface-dwelling organisms. More than half the gene groups scientists found in this water come from fungi, so scientists hope that they will include genes for new antibiotics.

>> **Make sure everyone in the world has access to high-quality antibiotics.** Bacteria do not respect national borders. As people and goods travel the world, they take their bacteria with them. If people in resource-poor countries don't have access to effective antibiotics, then their infections have a chance to multiply and pass on any resistance genes they have. This situation can be made worse if people try to treat resistant infections with drugs that no longer work. All this does is speed up evolution. Unfortunately, the political will to make sure all people have access to effective medicine often fails in the face of economic priorities.

Exploring Traditional Uses of Fungi

Cultures around the world have used fungi as medicine for thousands of years. Traditional practitioners in China, India, Africa, and Russia used observation and intuition to develop medicines and practices based on the relationships between people and nature. Most medicines were prepared from plants and fungi. The knowledge of which fungi to use, how to prepare them, and what to use them for was passed down from generation to generation. In cultures that had written language, like the Chinese, these practices were written down, but in others, the knowledge was passed verbally. Although traditional practices weren't tested using the scientific method, they were tested by repeated use over many years.

REMEMBER

Traditional medicine is the sum of the knowledge, skills, and practices based on the theories, beliefs, and experiences indigenous to different cultures, whether explicable or not, used in the maintenance of health and the prevention, diagnosis, improvement, or treatment of physical and mental illness (definition from World Health Organization).

Some medicinal mushrooms are shown in Figure 14-3. Many of these are known around the world today, including:

>> **Turkey tail** (*Trametes versicolor*), or yún zhī, is a polypore fungus used in China and Japan to stimulate the immune system. The fungus can be used directly to make a tea, or it can be dried and ground into a powder.

>> **Cordyceps** (*Ophiocordyceps sinensis* and *Cordyceps militaris*), or dōng chóng xià cǎo, is a parasitic fungus used in traditional Chinese medicine to support kidney and lung health. Infected caterpillars are dried and used to prepare a tea.

>> **Reishi** (*Ganoderma lingzhi*), or líng zhī, is a polypore fungus used in China, Japan, Korea, and other Asian countries as a tonic and to treat various diseases, including chronic liver disease, high blood pressure, insomnia, bronchitis, gastric ulcer, diabetes, and cancer. It's available as powders, dietary supplements, and tea, which are produced from different parts of the mushroom, including mycelia, spores, and the fruiting body.

>> **Shiitake** (*Lentinula edodes*), or xiāng xùn, is a mushroom that's been eaten as a healthy food and used for thousands of years in Japan, China, and Korea to improve health and longevity, as well as improve circulation. They are used as a whole food and can also be found as a powder in supplements.

>> **Chaga** (*Inonotus obliquus*), or чáра, is a conk that grows on birch trees. It's been used in Russia since the 12th century as a treatment for gastrointestinal disorders, cardiovascular diseases, diabetes, and even cancer. Traditionally, it's dried, ground into a powder, then used to make a hot beverage.

>> **Lion's mane** (*Hericium erinaceus*), or hóu tóu gū, is a functional food that has a mild flavor and nice texture. They are eaten raw, dried, or cooked. It's used medicinally by Native Americans and in traditional Chinese medicine for infections, anxiety, stress, and depression.

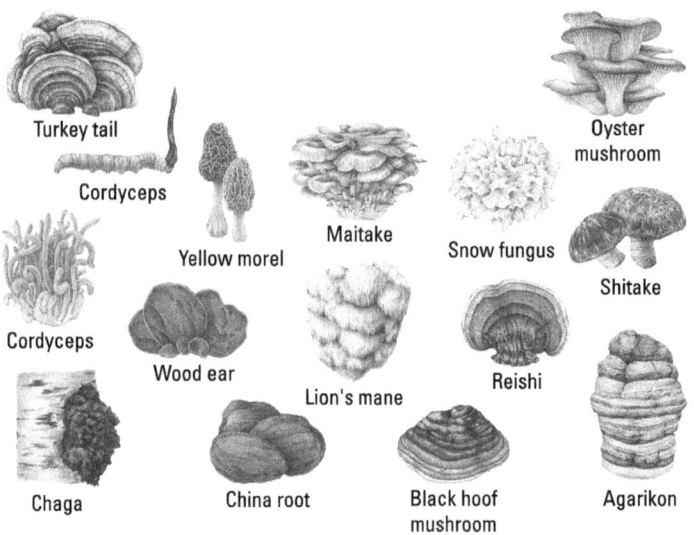

anitapol/Adobe Stock Photo

FIGURE 14-3: Medicinal mushrooms.

Anthropologists and scientists collect knowledge of traditional practices around the world and share it within their own cultures.

Ethnomycology is the study of how people traditionally use fungi for food, medicine, and cultural practices.

REMEMBER

Traditional medicine is typically received with skepticism by Western medicine that's based on the scientific method. Although both approaches share features such as observation and repetition, Western medicine has additional criteria for evaluating whether a treatment is effective:

>> Traditional medicine uses whole parts of plants or mushrooms, whereas Western science seeks to isolate one specific chemical component from the plant or mushroom.

>> Although practitioners may make traditional medicines according to time-honored traditions, the exact strength of a medicine may vary due to a variety of factors. When pharmaceutical companies produce a medicine, they quantify the exact dose of the active ingredient.

>> Traditional medicine is given to everyone who needs it and is judged based on how well it works through generational use. Scientific medicine requires a *controlled experiment*, where some subjects are given the medicine while others are given a false medicine called a *placebo*. A medicine is judged effective only if the people who receive the real medicine have a better outcome than those who receive the placebo.

>> Traditional knowledge is gathered and conveyed through the relationship between people and the natural world and often includes a spiritual component. Western science, on the other hand, claims objectivity and excludes spirituality from scientific practice.

At least partially because of these differences, Western practitioners historically dismissed traditional medicines. Recently, that seems to be changing. Some Western scientists show more respect for traditional knowledge and some medical practitioners in Western nations are even using traditional medicine.

REMEMBER

Alternative, integrative, and *complementary medicine* all refer to medical practices that include treatments that haven't been tested by the scientific method.

At the same time, alternative medical practices are growing, and Western scientists are using their methods to test plants and fungi used in traditional medicine to try and verify whether they work.

Ultimately, people's health would probably benefit from combining the best of both worlds. The Mi'kmaw people of the Northeastern North America have a word for this viewpoint, "etauptmumk," which means "two-eyed seeing." They say that we must look through one eye at tradition and through the other eye at modern approaches.

Moving from Traditional Knowledge to Western Medicine

Both traditional knowledge and Western medicine have their benefits in the treatment of disease. Traditional knowledge is powerful and available to people who might not have access to Western medicine. Western medicines are precisely formulated and proven to work in controlled studies. The greatest overall benefit to people may occur if we can learn from our ancestors and also test their medicines with Western approaches.

The path from traditional knowledge to Western medicine is a slow and careful one. Although there isn't just one way to walk this path, many drugs have been developed from a similar series of steps:

1. **Explore traditional knowledge to learn about medicinal plants and fungi and how they are traditionally prepared and used.** This should be done respectfully and in relationship with the people who hold the knowledge so that there are positive outcomes for all parties. Unfortunately, many examples exist where knowledge has been taken without reciprocity and where people and the environment have been harmed as a result.

2. **Identify the chemical compounds in the plants or fungi.** Scientists grind up tissues, then remove or extract chemicals in different solvents. After extraction, the chemicals in the solutions are separated, or *purified*, and then identified based on their chemical properties.

3. **Test individual chemicals on cell lines in the laboratory.** If scientists want to test the ability of a chemical for its potential to cure cancer, they might test it on isolated cancer cells growing in the lab. Likewise, if they want to see if a chemical affects the immune system, they might test how the chemical changes the behavior of white blood cells growing in cell culture. Testing chemicals on cells growing in dishes in the laboratory is a good first step to screen all the potential chemicals isolated from a biological specimen without any risk to people.

4. **Test promising chemicals for their ability to cure disease in mice.** Mice are mammals and have many characteristics in common with people. They are often used as model organisms for research that might benefit human health. Scientists can cause diseases like cancer and diabetes in mice and then test potential treatments. The potential benefit to humans is generally regarded as more important than the harm to mice. If a chemical can reduce disease in mice, it may then get approved for clinical trials in people.

5. **Test beneficial chemicals in people.** In Western nations, this type of testing typically requires government approval. Scientists must apply for permission to conduct these experiments, which are called *clinical trials.* They must show evidence from past experiments to justify the potential benefit of the treatment. Volunteers in clinical trials must be informed of what will be done and of all possible risks. If negative outcomes occur during the trial, it must be stopped immediately. If a chemical positively impacts people's health and does not have severe side effects, it may eventually be approved as a medicine.

REMEMBER

For most medicinal benefits from traditional knowledge about fungi, the benefits have not yet been fully tested via Western approaches. Although some fungal chemicals have shown activity in the laboratory in cell lines or mice, very few have been tested in people.

WARNING

Many companies sell products that are advertised as being good for your health because they contain mushrooms. Currently, however, there aren't any regulations about what can be officially called a mushroom. Traditional medicinal use of a mushroom almost always involves using the actual fruiting body, which is where most of the active chemicals tend to be. Some companies, however, are using mycelium to create their mushroom products. In some cases, substrates such as grain may go into the product along with the mycelium. If you buy mushroom products hoping for health benefits, be sure to research your products thoroughly. Buying or growing the mushrooms themselves may be your most reliable option.

Fighting cancer

Cancer occurs when normal body cells become mutated so that they forget their original job in the body. They start dividing when they aren't supposed to, forming lumps of cells called *tumors*. As they continue to divide, they mutate further, and may leave their tissue of origin, finding their way to the blood or the lymph and spreading throughout the body.

Wherever cancer cells land, they continue to divide and form tumors, disrupting organ function. They can release chemicals that attract blood vessels to grow toward them and supply them with the food they need to keep growing. If they aren't stopped, the tumors ultimately lead to organ failure and death.

Treating the disease

Doctors choose cancer treatments based on the type of cancer, meaning which type of cells originally mutated, and on the stage of the cancer. When cancer is caught early, before metastasis occurs, local treatments such as surgery and radiation can be effective. As long as doctors can remove all of the cancer cells by surgery or kill them with radiation, the patient can be cured. If cancer has already spread, then doctors need to use treatments that also spread throughout the body, such as anticancer drugs.

Scientists are investigating mushrooms traditionally used to prevent or treat cancer to see if they contain cancer-fighting chemicals. Typically, scientists extract the chemicals and test them on cancer cells growing in the lab. Chemicals can also be tested against cancer in lab animals like mice. If the chemicals show promise in laboratory experiments, then they move on to clinical trials in human cancer patients.

Targeting tumors

Turkey tail, *cordyceps*, reishi, chaga, and lion's mane all make compounds that show antitumor activity:

>> Turkey tail produces a molecule called polysaccharopeptide krestin (PSK) that has antitumor activity. Patients who received PSK along with conventional cancer medicine had better outcomes than people who received conventional cancer treatments only.

>> *Cordyceps* produces a molecule called cordycepin. In lab experiments, cordycepin inhibited the development of liver cancer cells. Extracts from *C. militaris* caused cell suicide in both human lung cancer cells and leukemia cell lines.

>> Reishi produces polysaccharides and chemicals called triterpenes that have anti-cancer effects. In laboratory experiments, extracts from the mushroom stopped several types of cancer cells from growing, including cancer cells from human lung, liver, breast, prostate, cervix, ovary, and bladder. In many cases, extracts also triggered cell suicide in the cancer cells. In other experiments, a small protein from the mushroom, called GL-PP, was able to stop the recruitment of blood vessels to cancer cells in culture.

>> Chaga produces chemicals called triterpenoids that have the ability to inhibit the development of cancer in a common mouse skin assay and to promote cell death in cultured lung cancer cells.

>> Lion's mane produces a diterpenoid called erinacine A that triggers cell death in cultured gastric carcinoma cells. This compound also killed colon cancer cells in live mice.

Promoting cardiovascular health

Cardiovascular disease (CVD) is the leading cause of death in the world, killing almost 18 million people per year. Most deaths due to CVD come from cardiac arrest (heart attack) and stroke. The biggest risks for CVD result from lifestyle choices such as eating an unhealthy diet, not exercising, smoking, and drinking too much alcohol.

Hardening the arteries

The cholesterol in your blood has a big impact on whether you'll develop cardiovascular disease. Cholesterol in your blood isn't automatically bad because you need cholesterol for your cell membranes and hormones such as testosterone and estrogen. Little particles of fat and protein called lipoproteins carry cholesterol around your body. The type of lipoproteins in your blood indicates whether you are at risk for CVD:

>> *High-density lipoproteins* (HDL) carry cholesterol from your cells and arteries to your liver. HDL is sometimes referred to

as "the good cholesterol." According to the Cleveland Clinic, your HDL number should be high (above 60).

>> *Low-density lipoproteins* (LDL) carry cholesterol from your liver to your cells. LDL is sometimes referred to as "the bad cholesterol." According to the Cleveland Clinic, your LDL number should be below 100.

Doctors may use two different measures to calculate your risk for CVD:

>> **Ratio of total cholesterol to HDL:** To calculate this number, divide your total cholesterol by your HDL. The higher the number, the greater the risk. According to the University of Rochester Medical Center, your ratio should be lower than 5 to 1, and a ratio below 3.5 to 1 is considered very good.

>> **The amount of non-HDL cholesterol:** To calculate this number, subtract your HDL from your total cholesterol. According to the Mayo Clinic, the optimal level of non-HDL cholesterol for most people is less than 130 milligrams per deciliter (mg/dL), which is 3.37 millimoles per liter (mmol/L).

High levels of non–HDL cholesterol are dangerous because they trigger damage to your blood vessels. When you have too many LDL particles, they can enter the lining of your arteries and accumulate. When these lipids begin to break down (oxidize), they damage your arteries, triggering an immune response. Your immune system walls off the damage by forming plaques that include white blood cells. These plaques can ultimately lead to clots that block the flow of blood. Lack of blood flow to the heart causes heart attacks, while lack of blood flow to the brain causes strokes.

Controlling cholesterol with statins

In the 1960s, a Japanese scientist named Akira Endo turned his expertise in fungal chemistry toward the problem of cholesterol. Other scientists had recently discovered a key enzyme required for the synthesis of cholesterol, and Endo wondered if any fungi produced a molecule that might block the cholesterol-synthesizing enzyme.

Endo and a colleague, Masao Kuroda, began the lengthy task of screening fungal strains. After testing 6,000 strains of fungi, they discovered that *Penicillium citrinum* produced a compound that blocked the necessary enzyme, lowering cholesterol synthesis. Ultimately, they named the fungal compound mevastatin and began testing its ability to lower cholesterol in dogs.

Other companies and scientists began searching for similar drugs. The pharmaceutical company Merck discovered lovastatin produced by the mold *Aspergillus terreus*. When Merck tested its drug in people who had an inherited form of extremely high cholesterol, the drug successfully lowered LDL cholesterol and produced very few side effects. More testing and drug modifications followed, eventually resulting in the ability to synthesize statin drugs rather than isolating them from fungi. The drug company Parke-Davis produced one of the most successful of these synthetic statins, which they named atorvastatin. You may have heard of it by its trade name, Lipitor.

The availability of statin drugs had an impressive impact on cardiovascular disease. In the United States alone, the number of deaths from CVD decreased by about 25 percent since 1994. This wasn't due to lifestyle changes, but rather to the availability of medicines, like statins, to protect cardiovascular health.

Controlling diabetes

Diabetes is on the rise globally, increasing from 200 million affected people in 1990 to 830 million in 2022. The highest increases were in low- to middle-income countries, and over half of the affected people didn't take medicine to control their diabetes in 2022. Uncontrolled diabetes can lead to blindness, kidney failure, heart attack, stroke, and the need to amputate lower limbs due to poor circulation.

Defining diabetes

Diabetes is a disease that results from high blood sugar. Your body has systems that normally regulate the amount of sugar in your blood. When you eat a meal, your digestive system breaks the large molecules in your food into smaller molecules. The cells that line your small intestine absorb those molecules, which include glucose, into your blood so they can circulate around your body and reach all of your cells.

When your body detects that glucose is available, your pancreas releases the hormone insulin. Insulin also circulates around your body, attaching to receptors on your cells to let them know glucose is available. Your cells absorb the glucose and use it as a source of energy.

REMEMBER

Blood glucose goes up after a meal, triggering the release of insulin from the pancreas. Insulin signals cells to take glucose out of the blood, so blood glucose falls again.

Diabetes results when this system isn't working properly. Two main types of diabetes result from two different causes:

>> **Type I Diabetes** results from an inability to produce enough insulin. Scientists think this damage results from autoimmune disorders, where the body's immune system attacks itself.

>> **Type II Diabetes** results from an inability to respond to glucose. This type of diabetes is also sometimes referred to as insulin resistance, because the cells of the body don't listen to the signal from insulin. Risk factors for this type of diabetes include obesity, lack of exercise, age, and genetics (race and family history).

Influencing gut microbes

Mushrooms are an excellent source of diverse *prebiotics* that influence the growth of gut microbes (head back to Chapter 12 for more on prebiotics). Several studies in mice have shown that eating the polysaccharides in mushrooms can alter the gut microbiota in ways that may prevent the development of Type II diabetes. In one study, scientists fed mice a daily serving of white button mushrooms (*Agaricus bisporus*) that would be equivalent to three ounces per day for a human. The mice that ate their daily mushrooms experienced shifts in their gut microbiota, with increases in bacteria that produced more of the short-chain fatty acids succinate and propionate. Scientists previously demonstrated that these two fatty acids have a positive impact on the control of glucose in the body.

TIP

Mushrooms are also a great dietary choice for diabetic individuals because of their cholesterol, fat, and carbohydrate content, and high protein, mineral, and vitamin content.

Lowering blood sugar

In addition to the prebiotic impact of mushroom fibers, fungi also contain specific chemicals that may alter blood glucose. Scientists researching the potential medicinal impacts of fungi on diabetes have seen many positive results:

>> *Cordyceps militaris* extracts lowered blood glucose in diabetic rats and mice. In rats, extracts actually increased glucose metabolism while also decreasing insulin resistance.

>> Polysaccharides from **reishi** lowered blood glucose in diabetic mice. In addition, the triterpenoid molecule ganoderan B increased the amount of insulin in the blood of diabetic mice. Scientists also set up a study with 71 adult patients with Type II diabetes. Half of the patients were given supplements of polysaccharides from reishi three times a day for 12 weeks, while the other half received a placebo. The blood glucose of patients who received the polysaccharides decreased significantly relative to those who were given the placebo, as did their levels of glycosylated hemoglobin (HbA1C), which is a good indicator of the amount of sugar in the blood over time.

>> **Shiitake** mushrooms can help protect pancreatic beta cells, boost insulin production, and lower blood glucose levels. In one experiment, scientists gave diabetic rats an extract from shiitake mushrooms and found that it lowered their blood glucose and increased insulin production.

>> Extracts from **lion's mane mushrooms** also lowered blood glucose levels and increased insulin production in diabetic rats.

Preventing Dementia

The elderly population is growing rapidly in many countries, including China and India, leading to estimates that 80 to 90 million people will be over the age of 65 in 2050. As people age, they develop chronic conditions like dementia. In 2020, 55 million people in the world were living with dementia, and this number is predicted to reach 139 million by 2050.

REMEMBER

Dementia is a term for a decline in memory, reasoning, or other thinking skills. *Alzheimer's disease* is the name of a specific brain disease that causes 60 to 80 percent of dementia cases.

Defining Alzheimer's

The signature signs of Alzheimer's disease are the death of neurons in the forebrain, hippocampus, and cerebral cortex of the brain, along with protein deposits called amyloid plaques. The protein in the plaques, called Amyloid β protein, helps trigger the changes in the brain, including the death of neurons. The drugs currently available to treat Alzheimer's help slow the progression of the disease, but they don't reverse the damage to cognitive functions.

Stimulating nerves

Compounds in fungi have shown great promise for the protection of neurons and the prevention of dementia:

>> When scientists gave rats extracts from *Cordyceps ophioglossoides*, then exposed them to chemicals that would normally trigger the death of neurons, they found that rats given the extracts showed less cell death than rats that received a placebo. The rats that received the extract also performed better in a maze challenge, indicating that they experienced less loss of spatial memory.

>> Extracts from **reishi** stimulated the differentiation of neurons in rats and prevented neuron cell death.

>> Scientific studies with **lion's mane** indicate that it has great potential for promoting neurological health. Extracts from this fungus stimulated the differentiation and growth of neurons and helped minimize the impact of damage due to strokes in mice. Scientists conducted a study on 50- to 80-year-old Japanese men and women with mild cognitive impairment. Half of the participants took supplements that contained powdered lion's mane three times per day, while the other half took a placebo. All participants took a test designed to measure dementia before and after the experiment. The scientists found that cognitive function improved in the subjects who took the lion's mane supplements, and that the scores increased with the duration of intake.

IN THIS CHAPTER

» **Identifying the types of mushrooms that are easy to grow**

» **Choosing the best method for growing mushrooms at home**

» **Taking a tour of mushroom farming methods**

Chapter **15**

Farming Fungi

M ushrooms are nutritional powerhouses, providing umami flavor along with many health benefits. If the mushrooms you want to incorporate into your diet aren't available at your local market, you could consider trying to grow them at home. Growing mushrooms indoors doesn't take up much space, and it can be a lot of fun. If you grow mushrooms outside, you can improve your soil and the growth of your plants. In this chapter, you find out which mushrooms are easiest to grow at home and explore which methods would work best for you.

Growing Mushrooms at Home

Growing mushrooms at home is a great way to add more of the nutritional and medicinal benefits of fungi to your life. It doesn't require much space, and you can grow varieties that aren't typically available in your local market. If you aren't confident about foraging wild fungi, it's a great way to add some diversity to your culinary repertoire. It's easy to begin with ready-to-go kits, and you can expand into more extensive setups if you find you like mushroom farming.

Choosing a mushroom

Some mushrooms are easier to grow than others. Some of the easiest to grow at home are oysters, shiitake, lion's mane, and pioppino mushrooms (shown in Figure 15-1). Of these, oyster mushrooms are great for beginners because they grow rapidly on many different substrates, including coffee grounds. They can outgrow potential competitors in the culture, like contaminating molds, so they require less work from you. People also commonly grow enoki, wine cap, reishi, maitake (hen of the woods), brown beech mushrooms, and royal trumpets.

REMEMBER

Mycelium grows on materials called *substrate*. Some mushrooms can grow on a wide variety of substrates, while others have more specific needs.

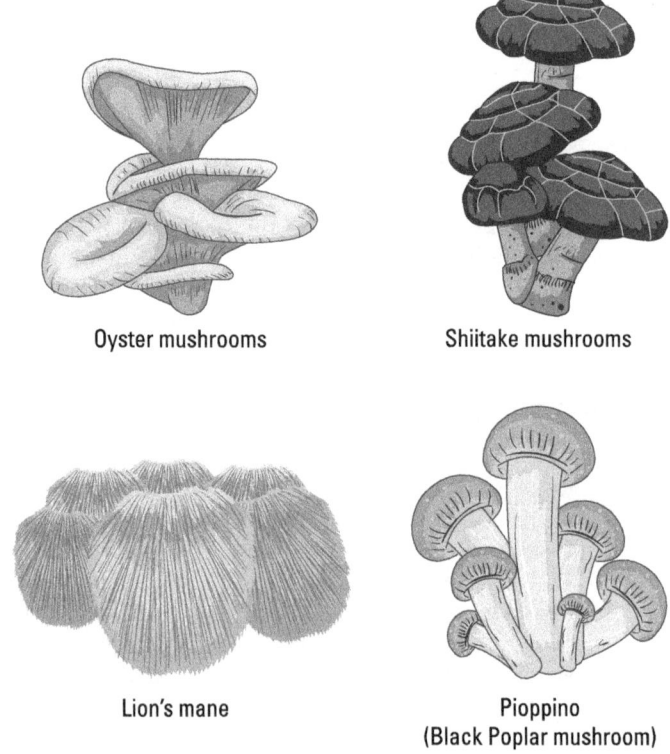

Oyster mushrooms   Shiitake mushrooms

Lion's mane   Pioppino
                                          (Black Poplar mushroom)

FIGURE 15-1: Easy-to-grow mushrooms.

Deciding on your method

When choosing how you want to grow mushrooms, you need to think about how easy you want it to be, whether you want to grow indoors or outdoors, and how much you want to spend. Some of the more common methods are shown in Figure 15-2.

Whichever method you choose, your mushroom will go through two growth phases:

>> **The *spawn run* where the mycelium grows through the substrate.** If you buy a kit that includes a mushroom block, the block will already be inoculated with mycelia, so your mycelium will move to the fruiting stage fairly quickly (as little as ten days). If you inoculate a substrate yourself, it will take longer. Outdoor log inoculations can take months to years to complete this phase.

>> ***Fruiting* occurs when the mycelium produces mushrooms.** You'll know your mycelium is beginning to fruit when you see little knots of mycelium, called *primordia*, beginning to form. This stage is called *pinning* because the primordia look like little pins. This is a good time to make sure your mycelium has optimal growing conditions for producing mushrooms:

  ● *Light:* Mushrooms aren't plants, so they don't need light. Most species do best in indirect light. The button mushroom, *Agaricus bisporus*, prefers the dark.

  ● *Temperature:* Preferred temperatures depend on the species, but many mushrooms grow well between 55 and 70 degrees F. The best temperature for the production of mushrooms may be slightly lower, and some species need a temperature drop to trigger good mushroom production.

  ● *Moisture:* Mushrooms like high humidity, with most species preferring 80 to 95 percent. You can achieve this level of humidity by enclosing your mycelium in a ventilated space and misting it regularly. Once your mycelium starts to make mushrooms, you may need to adjust the humidity to be slightly lower.

  ● *Air flow:* Your mycelium needs good air exchange. Fungi take in oxygen gas and release carbon dioxide just like people. They need air movement to remove CO_2 and bring in fresh O_2.

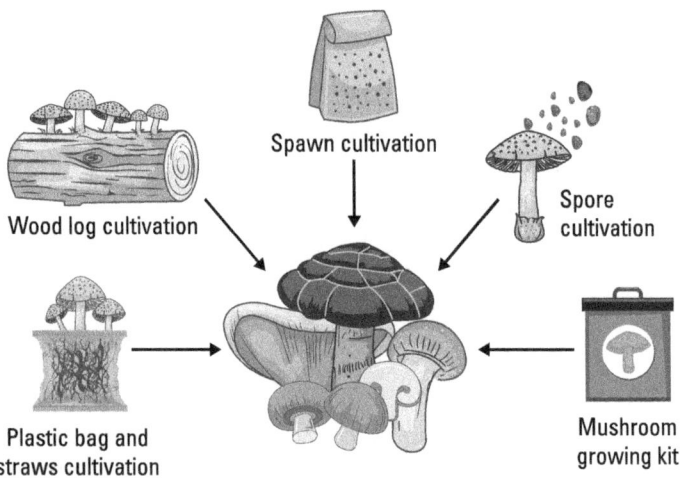

FIGURE 15-2: Some methods for home cultivation.

Grow indoors with mushroom grow kits

This is by far the easiest method, and it doesn't require a huge investment of money or space. For around $30, you will receive a box that contains a pre-inoculated block of substrate. Most kits also include a spray bottle for misting your mycelium. The kits come with instructions for what to do, but the basic protocol is to cut open the plastic package around the block and mist it twice a day. You get to watch your mushrooms grow, and you should be eating them in just a few weeks. When the block is done growing, you can buy replacement blocks for around $20. Many companies offer guarantees on their kits and have helpful staff, so it's worth shopping around a little (see Chapter 18 for information on some reputable companies).

Grow indoors in a plastic bag, bucket, or bin

This method requires a larger investment of money to get set up and requires more preparation, but it can provide larger yields. You can buy kits that include substrate and a container, or you can assemble the materials yourself. If you purchase a kit and a spore culture, you could get set up for between $50 and $100. If you buy a fancy kit that has filters to keep out contaminants and automatic misters, you could spend $300. No matter which container you choose, the basic method is this:

1. Sanitize your equipment to eliminate contaminating mold spores. Depending upon the equipment, you can sanitize with boiling water or 70 percent rubbing alcohol (isopropyl alcohol).

2. Fill your container with substrate.

3. Inoculate your substrate with spawn. Depending upon what materials you buy, you might have to create a grain spawn by inoculating spores onto a nutrient-rich grain first.

4. Mist and grow.

Spawn is a mixture of spores and substrate.

Grow indoors in a growth chamber

If you like growing mushrooms and want to grow multiple varieties at once, you may consider building or investing in a growth chamber such as a Martha Tent. A kit for just a greenhouse shelf can cost as little as $50. If you want to buy a kit for a Martha Tent with built-in misters and fans, it could cost as much as $450. If you are willing to start with the low-priced greenhouse shelf and add your own humidifier and fans, you can probably build a decent grow chamber for $100 to $150.

Grow outdoors on logs

You can purchase logs that are already set up and ready to go, or you can make the logs yourself. If you want to purchase an inoculated log, you'll probably spend about $35 for a small log. For about the same amount of money, you could buy a kit that has everything but the log itself. This would require more work, but you'd be able to inoculate a larger log. Be sure to choose a fresh log, not one that is already decomposing.

The first thing you'll need to do is figure out what type of wood your mushroom will grow on. You can easily look this up on the internet, but here are a few of the most common pairings to get you started:

>> **Shiitake** mushrooms' favorite wood is beech or oak, but they will also grow on alder, birch, cherry, poplar, and several other species.

>> **Oyster** mushrooms love aspen, cottonwood, and willow, but they will also grow on alder, beech, birch, elm, and many other substrates.

>> **Lion's mane** loves beech and hard maple, but will also grow on alder, aspen, birch, willow, and others.

>> **Maitake** loves oak, but will also grow on alder, beech, cottonwood, and more.

The recommended size for a mushroom log is somewhere between three and eight inches wide and three to four feet long. You will also need to either purchase or grow some spawn of your chosen mushroom.

Once you have your wood and spawn, these are the basic steps:

1. **Drill holes in the log that are about one inch deep and four to six inches apart.**

   Alternate each row of holes so that you create a diamond pattern.

2. **Fill your holes with spawn.**

   You can use loose spawn, like shiitake inoculated into sawdust, or you can buy plugs, which are small wooden dowels that were inoculated with mycelium.

3. **Seal your holes and the ends of the log with melted wax.**

4. **Put your log outside in the shade.**

   If you have a stretch of dry weather, it's good to mist the log periodically.

Grow outdoors in garden beds

If you already have a vegetable or flower garden, this method may make the most sense for you. Not only would you have the benefit of another type of food, but the mushrooms will help break down organic matter and supply more nutrients to your plants.

You can add mushrooms to your existing gardens by mulching with substrates that support mushroom growth. The general method is to put down a layer of substrate, then scatter spawn onto the substrate, then add another layer of substrate. You can purchase a bag of spawn that will inoculate 16 square feet for about $30.

TIP

If you have paths around or through your garden, you can put down a layer of cardboard before you add your substrate and spawn. Cardboard from the United States, Canada, and Europe should be free from dioxins and safe to use in the garden. The cardboard will break down slowly and help block weeds.

You need to choose a compatible substrate-mushroom combination for your garden beds. Wood chips make a good substrate for garden paths and around trees and bushes. Wine cap mushrooms (*Stropharia rugosoannulata*) grow well on several types of wood chips. Straw is a good substrate for mulching vegetables and also for the growth of various oyster mushrooms.

Looking at Commercial Production of Fungi

The commercial production of fungi involves the same basic steps as growing fungi at home. Farmers inoculate substrate with spawn, allow the spawn to run through the substrate, then adjust environmental conditions to optimize the production of mushrooms. Because mushroom farmers work with such large volumes of material, and because they rely on their product for income, they have to be extremely careful to prevent contamination of their crop by other fungi.

Starting life in the laboratory

Some commercial farms maintain their own culture collection of fungi that they grow. Fungal mycelia of individual species can be kept for long periods of time under refrigeration in test tubes that contain suitable nutrients. When a farmer is ready to start a new crop of mushrooms, they can transfer a bit of mycelium from a test tube onto dishes that contain nutrients thickened with agar.

Agar is a polysaccharide from algae that's used to thicken growth media in laboratories, just like gelatin can be used to thicken food. The growth media are poured into test tubes and dishes called Petri plates when it is still warm, then it's allowed to cool into a semi-solid surface that is excellent for growing microbes like fungi.

Once the mycelia from the culture collection are introduced into fresh plates of nutrient agar, they will begin to grow. The farmer can inspect the cultures and choose the ones that appear to be fast growers. They can also eliminate any cultures that show signs of contamination.

Some commercial farms don't maintain their own cultures, preferring instead to order spawn for each new crop.

Getting a boost as grain spawn

Farmers increase the amount of mycelium available to inoculate substrate by first transferring it to sterilized grains such as millet, rye, or wheat. A small piece of mycelium can be cut from the Petri dish and added to a sterilized jar of grain. The jar is shaken to disperse bits of hyphae throughout the nutrient-dense grain, which supports the rapid growth of the mycelium. The jar is placed at a good growing temperature for the mycelium and allowed to grow until the mycelium is spread throughout the grain.

Preparing the substrate

While the mycelium is growing on the grain, the farmer prepares the substrate for inoculation. The type of substrate varies depending on the species of mushroom. Some common substrates include sawdust, compost, straw, and coffee grounds. In most cases, the substrate must be sterilized to kill all other microbes before it is inoculated with the desired mycelium. Commercial farmers have large pieces of equipment similar to pressure cookers that can heat-sterilize their substrates.

Once the substrate is ready, it's loaded into bags or trays. In the case of bag cultivation, many growers load their substrate into heat-resistant plastic bags, then sterilize the substrate directly in the bags. The bags have filtered openings in them to allow gas exchange for the growing fungus without allowing contaminants to enter the bag.

Inoculating the substrate

Farmers use the grain spawn to inoculate the substrate. For most cultivated mushrooms, the potential for contamination by other fungi is a concern, so the grain spawn is carefully inspected for signs of undesirable mycelium. The transfer of grain spawn to bagged substrate is carefully done in front of an air filter that helps prevent contamination during the transfer step.

Bags

Farmers incubate their bags of inoculated substrate in conditions that are optimal for mycelial growth. When the mycelium has grown through the substrate and started to form primordia, the farmer cuts the bag to allow increased oxygen to the mycelium and to allow the mushrooms space to expand out of the bag. If mushroom formation requires different environmental conditions than mycelial growth, the farmer can move the bags to a new growth chamber.

Trays

The tray method is typically used for the cultivation of the white button mushroom, *Agaricus bisporus*. Once the sterilized substrate is loaded into trays and inoculated with grain spawn, the farmer monitors the growth of the mycelium through the substrate. When pins start to form, the farmer adds another layer, called a casing, to the tray.

REMEMBER

A *casing* is a layer of low-nutrient material added to the top of the mushroom-growing substrate. The casing helps hold in moisture and create optimal conditions for mushroom production. Commonly used casing materials include peat moss, vermiculite, and the fiber from coconut husks, which is called coco coir.

Harvesting the crop

As clusters of mushrooms grow from the substrate, the farmer cuts them off with a knife so they can be packaged for sale.

REMEMBER

Each crop of mushrooms is called a *flush*.

Farmers typically get several flushes from each batch of substrate, but the number of mushrooms decreases with each flush, while the risk of contamination increases. Once a layer of substrate has stopped being productive, it can be composted.

The Part of Tens

IN THIS PART . . .

Explore the weirdest and wildest fungi in the world, including fungi that can turn ants into zombies, absorb radiation, and explode into a cloud of smoke.

Investigate ways fungi may be used in the future, from breaking down plastics to fighting infections to cleaning up pollutants in the environment.

Discover additional resources for expanding your knowledge about fungi.

Chapter **16**

Ten Fantastical Fungi Tricks

Fungi spend so much of their time invisible to the naked eye that people often don't notice the species literally growing under their feet. But every now and then, a fungus comes along that makes people take notice. In this chapter, you meet some of the weirdest and wildest fungi in the world.

Turning Ants into Zombies

If you're a fan of the TV series *The Last of Us,* then you've probably heard of "Cordyceps" — the fictional fungus that turns people into zombies driven to seek out and infect other people. That is science fiction, but a piece of science fact is that some species in the genus *Ophiocordyceps* can actually do just that — to ants! If a single spore lands on an ant, it can doom the ant to zombification.

Just like the human zombies in the TV show, the ants completely lose their identity and exist to serve the fungus. First, the fungal hyphae emerge from the spore and penetrate the ant's exoskeleton, spreading throughout the living ant and digesting it from the inside. The fungus takes control of the ant's movement and behavior, causing it to wander erratically and ignore the scent trails from its colony. Ultimately, the ant climbs up a stem to find a spot with the perfect conditions for the fungus to reproduce. When it locates the perfect height, light, and moisture conditions, the ant clamps its jaws in a death lock on a vein on the underside of a leaf. The fungus emerges from the corpse, sending up a fruiting body that releases new spores to fall on the ants marching below the plant so that the cycle can begin again.

Unlike the "Cordyceps" of science fiction, *Ophiocordyceps* doesn't penetrate the brains of its hosts. Instead, scientists think it controls the ants by directly manipulating their muscles. If the ants' brains haven't been taken over, it raises the question: How much are the ants aware of what is happening to them?

Attracting Insects with Rotting Meat

Imagine finding a gelatinous egg-shaped structure on the forest floor. Red pointed fingers begin to emerge, looking almost like an octopus unfurling its tentacles. The tentacles splay outward, extending to lengths of 7 centimeters, giving this creature the appearance of a teeny-tiny eldritch horror. Its smell, like rotting meat, fits right in with its looks.

Despite this otherworldly scenario, what you are imagining is a member of the stinkhorn fungi, commonly called the devil's fingers or octopus stinkhorn. Like other stinkhorns, it uses its delightful aroma to attract flies and other insects that like to lay their eggs in carrion. The insects eat the slime produced at the tips of the tentacles, picking up spores on the outsides of their bodies. When the insects move on to the next attractive meal, they spread the fungal spores to new areas.

Striking Fear in Mushroom Foragers

The destroying angel. The death cap. The deadly skullcap. Clearly, fungi with names like these are to be avoided. These fungi all produce chemicals called *amatoxins* that block your cells from making proteins, leading to cell damage, liver failure, and even death. A person who eats a toxic dose of these fungi won't notice right away, but about six hours later will experience intense abdominal distress, including nausea and watery diarrhea. They'll start to feel better over the next few days, but cellular damage will occur silently, eventually leading to liver and kidney failure.

Ninety-five percent of deaths due to eating poisonous mushrooms are caused by fungi that produce amatoxins. People sometimes confuse poisonous mushrooms for edible varieties, or they may eat them because they're looking for mind-altering effects.

WARNING

If you are going to forage for mushrooms, it's a good idea to familiarize yourself with the characteristics of some of these fungi, especially those with edible look-alikes. Most of the fungi that produce these toxins belong to the genus *Amanita*, but they're also found in other species, including some species of *Lepiota* and *Galerina*.

Absorbing Radiation

Fungi can grow in some pretty extreme environments, including those that receive high amounts of ultraviolet (UV) radiation, like in the mountains in the Arctic and Antarctic. These fungi are often dark brown or even black because they make melanin, the same pigment that darkens human skin and hair. Melanin protects the fungi from ionizing radiation, such as UV and gamma radiation. Fungi that can produce melanin and survive ionizing radiation are of interest to scientists, who wonder if these fungi could survive space travel or be used to create radiation barriers for people, such as astronauts or cancer patients undergoing radiation treatment.

Some fungi go beyond just resisting damage from ionizing radiation; they may actually use it as a source of energy. In terms of evolutionary innovations, this is huge. It could mean that these fungi act like plants, making their own food but using ionizing radiation instead of visible light. A group of scientists studying radiation-resistant fungi taken from Chernobyl, the site of the 1986 nuclear disaster, discovered that some of the fungi, like the basidiomycete *Cryptococcus neoformans*, not only grew toward sources of radiation but also grew faster when it was present. They named these strains *radiotrophic* fungi (*radio*=radiation, *troph*=feeding), and they're conducting research to try to figure out how the fungi convert the radiation to a usable form of energy.

Living Underwater

Before 2005, biologists would have said that gilled mushrooms only lived on land. After all, these mushrooms need air to provide them with oxygen for their metabolism and to disperse their spores. That's why Professor Robert Coffan was so surprised when he spotted what seemed to be a small brown mushroom growing completely submerged in the water of the Rogue River in Oregon.

Professor Coffan used a disposable underwater camera to take pictures of the little mushrooms to show to mycologists. He collaborated with Darlene Southworth and Jonathan Frank of Southern Oregon University, who quickly identified the mushrooms as members of the genus *Psathyrella*, but further identification became trickier after that because this genus contains hundreds of different species of little brown mushrooms that all look alike. The scientists spent years comparing the spores and DNA code of the underwater mushroom to that of other *Psathyrella* species before establishing that it was a newly identified species. They named it *Psathyrella aquatica.*

Many mysteries still surround this mushroom, including when and why its ancestors moved from land to the river. So far, scientists have only found the mushroom growing in a short stretch of the Rogue River, but they continue to search for it in other places. They're also trying to understand how it manages to live completely underwater. One hypothesis is that the large air bubble that gathers under the mushroom cap provides access to oxygen. Scientists also wonder how the mushroom manages to disperse its spores to the nearby habitat, and now the spores don't all wash away with the flowing river. Several types of insect larvae that graze on the mushrooms may help carry spores to nearby locations. The accidental discovery of such an unusual mushroom raises so many questions and illustrates how much we still have to discover among the fungi of the world.

Living Extra Large

Blue whales in the Antarctic can grow to over 100 feet in length and weigh up to 150 tons. The largest dinosaur fossils ever found belong to the plant-eating titanosaurs, which could grow to 122 feet long and are estimated to have weighed 70 tons. As big as these animals sound, though, the record for the world's largest organism goes to the soil fungus, *Armillaria ostoyae*. A genetically identical mycelium of this fungus grows through an area of 2,385 acres, which is over 100,000,000 square feet, in the Malheur National Forest in Oregon. Given its large area and typical weight, scientists estimate that this individual would clock in somewhere between 7,567 and 35,000 tons. It's no wonder that people have nicknamed this specimen "The Humongous Fungus."

Armillaria ostoyae — a type of root rot fungus — grows among the roots of coniferous trees, sometimes surviving on dead matter but also parasitizing living trees by penetrating their bark. The white sheets of mycelia remain hidden under the bark of infected trees, which may show symptoms of disease, such as discolored needles and resin oozing from the base of the trunk. As trees die, the fungus spreads outward to new trees via thick strands of hyphae called *rhizomorphs* (*rhizo*=root, *morph*=shape) that look a little bit like dark worms or thin branches. Cream to gold-colored "honey mushrooms" may be spotted growing around dead and infected trees in the fall after a rain.

As the fungus spreads outward from the initial site of infection, it creates vast circles of infection with old, dead trees and stumps at the center. One unexpected benefit of this destruction is that it opens up patches in the forest canopy, allowing for more diverse species to colonize the space.

Exploding into a Cloud of Smoke

The giant puffball, *Calvatia gigantea,* grows so large that from a distance, its round white shape may be mistaken for a soccer ball left on the field. The largest one ever recorded was over eight feet around and weighed over 48 pounds. That's so big that it may look more like a fluffy white sheep than a soccer ball. Puffballs are also edible, so a single giant puffball can make quite a feast.

Puffballs are basidiomycetes (Chapter 11), but instead of making spores on gills, they make spores in the interior of their round shapes. Mature puffballs have a small hole on the top through which spores shoot out in response to pressure from raindrops or walking feet (see Chapter 4 for more on spore dispersal). Scientists estimate that each giant puffball can release trillions of spores.

TIP

Puffballs are edible before they've started making spores. To check whether they're at the right stage, foragers slice the puffballs down the middle and look inside. If the interior is firm and solid white, then spore production hasn't started. If the inside shows discoloration to brown or yellowish colors, then the puffball is too old to eat.

WARNING

Slicing puffballs open also lets you check to make sure you don't have a different, possibly toxic fungus. Earthballs are basidiomycetes in the genus *Scleroderma* that also make spores on the inside of their ball-shaped bodies. These toxic fungi can be recognized by the dark color of their spores, which are visible as a dark purplish color when you slice them open. Some mushrooms, including the deadly *Amanita* (see "Striking Fear in Mushroom Foragers" earlier in this chapter), first emerge from the ground as little round buttons or eggs before the distinctive mushroom shape extends upward. If you slice open a potential puffball and see the outline of a mushroom shape or visible gills, you could have a deadly imposter.

Glowing in the Dark

Male fireflies flash their light to tell females that they're interested in mating. Marine creatures like squid and jellies may light up to distract predators, while other species like the anglerfish use light from symbiotic bacteria to attract prey. Although their reasons differ, all of these organisms rely on *bioluminescence* at some time in their lives.

Many bioluminescent organisms, such as these, make light by using an enzyme to trigger the reaction of chemicals called *luciferins* with oxygen, leading to the release of light energy as one of the products. The enzymes that make the reactions happen are called *luciferases* (*luc*=light, *fer*=bearing), so these chemicals are named for their role in producing light.

Some fungi use these same reactions to glow in the dark. One of these, the jack o'lantern mushroom (*Omphalotus olearius*), appears bright orange in the light but reveals its glowing green gills in the dark. Depending upon the species, different parts of bioluminescent fungi may glow. In some species of *Arrillaria* (see "Living Extra Large" earlier), the mycelium itself glows, leading to an eerie green glow from rotting wood in the forest at night. This glow, first described as "cold fire" by Aristotle over 2,000 years ago, is sometimes called foxfire or fairy fire today.

Scientists aren't entirely sure why some fungi glow in the dark, but they're working on figuring it out. One possibility is that the glow is just accidental, a byproduct of some other process. However, scientists in the forests of Brazil showed that insects were attracted to the glow of fake mushrooms made to glow like the real ones. The scientists speculate that the insects may come to feed on the mushrooms, picking up spores and helping the mushrooms to spread, much like bees visit flowers for nectar and take pollen away with them.

Growing in an Ant's Garden

You've probably heard about the huge asteroid that hit the Earth at the end of the Cretaceous period, causing catastrophic environmental changes that led to mass extinctions and wiped out most of the dinosaurs. Scientists think a huge ash cloud rose in the atmosphere, blocking sunlight and causing photosynthetic organisms and the animals that relied on them for food to die. In this apocalyptic situation, when organisms needed to innovate to survive, one group of ants figured out that fungi thrive in environments with no light and lots of dead things and decided to take up fungus farming. This tribe of ants, called the Attini, started collecting fungus in the wild and bringing it back to their nests to grow underground.

One group of Attini, the leaf-cutter ants, even developed a sophisticated farming system that includes the equivalents of fertilization and crop rotation. These ants live in forests of Central and South America and the southeastern United States, moving busily along paths they cut through the trees to find leaves that can serve as good substrate for their fungal crops. After they cut pieces from the leaves, they carry them aloft like green sails above tiny ant ships as they return to the garden chambers in their underground nests.

Once back at the nests, the ants place the leaf pieces in their fungal gardens where the fungi use their enzymes to break the leaves down for the nutrients they need to grow. The ants, who can't digest the leaves for themselves, feed on the fungi. Some of the fungal species even make special nutrient-rich structures called *gongylidia* for the ants to eat. When a fungal garden begins to get too old to be productive, the ants move bits of fungi into a fresh chamber and begin the cycle over again.

The relationship between leaf-cutter ants and basidiomycete fungi (Chapter 11) in the family *Lepiotaceae* became so interdependent over time that today, neither can survive without the other. Scientists think the symbiotic relationship between these species is mutualistic, meaning that both species benefit. Both fungi and ants have a secure food supply, and the ants carefully maintain the growing conditions for the fungi, clearing away competitive species of fungi and practicing good hygiene, like releasing their own waste away from the garden. Bacteria on the surface of the ants even play a role, producing enzymes that help in the digestion of leaf material and making antimicrobial substances that help keep potential pathogens out of the garden.

Fighting Cancer

The turkey tail fungus is a beautiful shelf fungus that grows in overlapping crescents on decaying wood. It gets its name from its undulating bands of color that resemble those on the feather displays of male turkeys. Turkey tails are polypore fungi, so they have little tunnels or pores on their undersides rather than gills.

Turkey tails, or *Trametes versicolor*, have been used as food and medicine for thousands of years. They have a long history in traditional Chinese medicine, where they're known as yun zhi and used as a tonic and to treat lung diseases. The Japanese call them *kawaratake* and use them to boost the immune system. Recent analyses have revealed that turkey tails are packed full of diverse chemicals that have potential as antivirals, antioxidants, and anti-aging compounds. They're also good for the liver and may help prevent diabetes.

Turkey tails are also known for their cancer-fighting abilities due to a complex carbohydrate called Polysaccharide-K (also known as PSK or krestin). Since the 1970s, Japanese doctors have given cancer patients PSK along with treatments like surgery or chemotherapy. In many studies on people suffering from various types of cancers, patients who received PSK along with other cancer treatments had better outcomes than patients who took a placebo instead of PSK. Some of the benefits of the addition of PSK were faster recovery of the immune system, less chance of cancer returning, and longer patient survival.

WARNING

People are understandably very excited about the potential anti-cancer properties of fungi. However, it's important to remember that research is ongoing. Mushrooms have many health benefits, but they haven't been proven to cure cancer by Western standards.

Chapter 17

Ten Ways We May Use Fungi in the Future

S cientists estimate that the number of fungal species on Earth is in the millions, but so far, we've only described about 150,000 species. Although we've studied only a small fraction of this group of organisms, we've already discovered that they have amazing potential to improve human quality of life and the health of the planet. In this chapter, you take a peek into a possible future where fungi help solve many of humanity's problems.

Breaking Down Plastics

Plastics are amazingly useful substances. They're light, flexible, and long-lasting, plus they're less expensive to manufacture than paper, glass, and metal. Ever since humans discovered how to make plastics, we've shaped them into an astounding array of products, from wire insulation and telephones to medical devices, bottles, and even clothing.

Unfortunately, the chemicals that give plastics their useful qualities also have downsides: They don't decompose quickly, and some of them are hazardous to the health of humans and other animals. These problems are especially troubling because plastic production continues to rise. Approximately 9,000 metric tons of plastic were produced between 1950 and 2015, leading to almost 7,000 metric tons of plastic waste. Currently, only 10 percent of plastic waste is recycled; the rest of the waste ends up in landfills and in the oceans.

The news today is increasingly filled with reports of the hazards of plastic waste to health and the environment. Plastic waste is responsible for 80 percent of the pollution in the ocean and can suffocate or injure marine animals. As plastics degrade, they can become small pieces called microplastics that enter animal bodies through air, water, or food intake.

Microplastics are small pieces of plastic less than 5 millimeters long (less than ⅛ inch).

Microplastics are now present in all areas of our environment, including all types of living things and throughout the human body. Although scientists are still studying the impacts of microplastics on health, they've already been observed to harm the digestion and health of agricultural animals, and studies in humans suggest they increase the chances of heart attack, stroke, or even death.

Clearly, we need to come up with strategies to better deal with our plastic waste. There are many ways we can address this problem. For example, we can

>> **Reduce** our use of the most problematic plastics like those found in shopping bags and beverage bottles.

>> **Recycle** more plastic and improve recycling practices to minimize the production of microplastics.

>> **Biodegrade** plastics using fungi or other organisms.

Scientists have identified more than 100 species of fungi that can degrade different types of plastics, including polyvinyl chloride (PVC) and polyurethane (PU). Most of the studies that show fungi can degrade plastic were done in the laboratory, so more work needs to be done to see if this would be useful on a larger scale.

Improving Biofuel Production

Currently, most countries produce biofuels primarily from feed-stocks that are high in sugar and starch because they are easier and less expensive to process. A major drawback to this approach is that it competes with the use of these crops as food. If pretreatment of woody materials could be improved, this would make non-food substrates like agricultural wastes, grasses, or sawdust much more attractive as fuel sources. These materials are cheap and plentiful and don't compete with the human food supply.

Scientists are investigating the potential of fungi to improve biofuel production, especially from woody materials. For example, they've tried using fungi that do white rot (see Chapter 2) to remove the lignin from the wood. Scientists are currently exploring other groups of fungi, ranging from those that live in the nests of leaf-cutter ants to endophytic fungi that live inside plants and seaweeds, hoping to discover new candidates to improve the efficiency of the pretreatment process.

Using Mycorrhizae as Natural Fertilizer

Modern agriculture relies heavily on the use of chemical fertilizers that can damage soil microorganisms, contribute to environmental pollution, and cause a decrease in water quality. An alternative way to increase plant nutrient uptake without harmful environmental effects is to use biofertilizers.

REMEMBER

Biofertilizers are products that contain single organisms or communities of organisms that are capable of colonizing soil around plant roots.

Chemical fertilizers directly provide nutrients such as nitrogen and phosphorus to plants, while biofertilizers introduce microorganisms that can help plants acquire these nutrients from the soil. Biofertilizers have several advantages over chemical fertilizers:

>> Biofertilizers don't contaminate water supplies.

>> The organisms in biofertilizers produce compounds such as enzymes and hormones that stimulate plant growth.

>> Biofertilizers are a renewable resource.

>> Biofertilizers are especially effective in dry climates because they enhance the water retention properties of soil.

>> Some biofertilizers suppress the growth of plant pathogens.

Arbuscular mycorrhizae (Chapter 3) are attractive candidates for biofertilizers because they form associations with 80 percent of plants, many of which are important food crops. One of the challenges of developing arbuscular mycorrhizae as biofertilizers is that they are obligate symbionts and can't be grown in culture by themselves. This makes it difficult to develop products on a large scale for agricultural use and is an area of active research.

Fighting Infections with New Fungal Compounds

Scottish doctor Alexander Fleming noticed the ability of penicillin to kill bacteria in 1928 when he saw a clear zone surrounding a fungus that had contaminated a plate of bacteria. The mold was *Penicillium*, and the clear zone was due to the killing power of penicillin against the bacterium. Not quite 20 years later, penicillin was being produced in large quantities and given to soldiers in World War II.

REMEMBER

In nature, bacteria and fungi compete with each other for resources. *Antibiotics*, chemicals produced by microbes to stop the growth of other microbes, are just one tool in their arsenal.

Since the discovery of the antibiotic properties of penicillin, scientists have screened many bacteria and fungi for their ability to produce antimicrobials. This led to the discovery of many antibiotics, including streptomycin and tetracycline produced by soil bacteria and cephalosporins produced by fungi.

Antibiotics gave people the upper hand on infection, enabling the treatment of many diseases that had often been death sentences in the past. Childbed fever, or *puerperal sepsis,* which had claimed the lives of up to 30 percent of new mothers, was effectively treated with penicillin. Penicillin also prevented the deaths of children due to scarlet fever. Other infections and diseases also met their match; it turns out even the bubonic plague bacterium was no match for antibiotics.

Today, we are realizing that even antibiotics have their limitations. The more we use them, the more we kill susceptible bacteria, leaving the more resistant strains to reproduce. When people get infected with these *antibiotic-resistant* strains of bacteria, we are sometimes back where we started, scrambling for ways to fight the infection.

Of course, doctors and scientists continue to try to find new ways to treat antibiotic-resistant bacteria. One way is to try to find new antibiotics that we've never used before. Only a relatively small percentage of fungi have been grown and studied in laboratories, so they represent a largely untapped potential for new drugs. To find new drugs, some scientists argue, we need to look in new places and at new species. This has led scientists to study fungi growing in very unusual conditions, like the radioactive environment around Chernobyl, in the hopes that these fungi will produce antibiotics we haven't seen before.

Controlling Biofilms in Medical Settings

You've probably encountered many biofilms in your life. If you've ever stepped on a slippery rock at the beach or washed out your pet's bowl to find it disturbingly slimy, you've encountered biofilms. People brush and floss their teeth in an effort to break up dental plaque, which is another example of a biofilm.

Biofilms are microbes living in a slimy matrix that's attached to a surface.

REMEMBER   Many microbial communities in nature exist in biofilms. The matrix that surrounds the microbes in the biofilm protects them from environmental stress. Most biofilms probably don't do you any harm, and biofilm communities perform many beneficial roles in the ecosystem. In the medical setting, however, biofilms can be a significant problem.

The protective matrix of a biofilm can prevent antibiotics and disinfectants from reaching potential pathogens. Biofilms that form in sinks or air conditioners may be hard to eradicate. Even more dangerous are biofilms that form on medical devices implanted into patients, such as catheters, pacemakers, or prosthetic joints. Even if a bacterium is sensitive to a particular antibiotic, doctors may not be able to kill it because the antibiotic can't get through the matrix to reach the bacterium.

Medical researchers are currently exploring the potential of fungi to help fight biofilms. Some fungi make compounds that help block the signals that tell bacteria to start making a biofilm. Other chemicals from fungi have been able to break down biofilms that have already formed. It may be that the best treatment in the future for an infection of a biofilm-forming bacterium will be a combination therapy of antibiotics plus a fungal metabolite that inhibits biofilm formation.

Improving Nerve Function

The prevalence of neurodegenerative diseases like Alzheimer's and Parkinson's diseases is increasing worldwide, especially in countries with aging populations. Neurodegenerative diseases like these are associated with the loss of nerve growth factors, proteins that promote the growth and survival of nerve cells in the peripheral sensory and sympathetic nervous systems.

Some medicinal mushrooms have a history of use in indigenous cultures for their ability to support the peripheral nervous system, for example, to increase alertness. When these mushrooms have been studied in the laboratory, the compounds they produce have stimulated the growth of neurons. Most of the neuroactive compounds discovered so far are from the lion's mane mushroom, *Hericium erinaceus*. This medicinal mushroom produces compounds that stimulate the production of nerve growth factors. Compounds like these may aid in the prevention or treatment of neurogenerative diseases in the future.

Eating More Fungal Foods

As awareness of the environmental and health impacts of meat production has grown, people have increasingly looked for alternatives. Many fungi stimulate the *umami* or savory taste receptors, making them a potentially satisfying substitute for the consumption of meat. Many vegetarian and vegan processed foods use mushrooms as the basis for their meat replacements. Because of the flavor, nutritional, and health benefits of mushrooms, the global market for edible fungi has grown rapidly in recent years, increasing from 419 to 899 million tons per year between 2000 and 2018.

The mushroom-growing industry has kept pace with this increase in demand. They've developed new and improved strains and automated many processes to make large-scale production of mushrooms more efficient. One area where the industry continues to grow is in the domestication of new species. Many fungi have yet to be identified, especially those growing in tropical areas. These fungi represent potential food sources that have yet to be explored.

Cleaning Up Pollutants

The human population on Earth is over 8 billion. As our population increases, so does the amount of waste we produce. Some wastes that are particularly dangerous and hard to clean up include oil spills, heavy metal contamination, radiation, polychlorinated biphenyls (PCBs), and polycyclic aromatic hydrocarbons (PAH). Clean water and land are essential to human health, so our survival as a species is potentially impacted by our ability to clean up these pollutants.

Mycoremediation uses fungi or fungal enzymes to clean up pollutants in the environment.

REMEMBER Using fungi to clean up pollutants has many advantages:

>> **Mycoremediation is relatively inexpensive compared to other conventional remediation.** Fungi are plentiful in nature and can be grown easily in the laboratory.

>> **Fungi are enzymatic powerhouses.** White rot fungi (Chapter 2) are particularly suited for mycoremediation because their lignin-degrading enzymes target many organic pollutants that are similar in structure to lignin, including toxic chemicals such as dichlorodiphenyltrichloroethane (DDT) and polychlorinated biphenyls (PCB).

>> **Many fungi are resistant to heavy metals.** Because they are resistant, they can grow in environments contaminated with heavy metal pollutants.

>> **Fungi can be grown in large quantities in bioreactors.**
Bioreactors are tanks that optimize the growth of microorganisms in large quantities. The pharmaceutical and sugar industries are already using fungi grown in bioreactors to break down their pollutants.

Fungi have demonstrated their ability to clean up a variety of pollutants. In one experiment, scientists treated diesel-soaked soil with three different cleanup methods: enzymes, bacteria, and oyster mushroom mycelium. After six weeks, only the soil treated with oyster mushrooms had reduced amounts of polycyclic aromatic hydrocarbons (PAH). In addition, the soil produced a healthy crop of oyster mushrooms. In another study, mycorrhizal fungi were able to reduce the accumulation of metals in wheat grown in zinc-contaminated soil. The cell walls of fungi can bind metals, allowing them to absorb contaminants, which then can be physically removed by harvesting the fungus.

Using Fungal Enzymes in Industry

Many industrial processes rely on the action of enzymes. The food industry uses enzymes to produce many foods, including cheese, juice, and corn syrup. The production of biofuels, pharmaceuticals, textiles, and detergents also requires enzymes.

REMEMBER

Enzymes are molecules, usually proteins, that speed up chemical reactions.

Fungi produce diverse and powerful enzymes capable of breaking down some of the most resistant molecules found in nature.

Fungal enzymes have many characteristics that make them useful for industrial processes:

>> Fungi are absorptive feeders (Chapter 2), so they release their enzymes outside of their cells. Fungi grown in bioreactors release large quantities of enzymes that are relatively easy to harvest.

>> Fungi produce a wide variety of enzymes useful in industrial processes. These include amylases for breaking down starch, cellulases for breaking down cellulose, lipases for breaking down lipids, and proteases for breaking down proteins.

>> Fungi can be grown on inexpensive substrates. This makes them a more affordable option for commercial purposes.

Farming for the Future

The growing knowledge of the importance of fungi to soil health and plant growth has contributed to the development of the field of *regenerative agriculture.*

REMEMBER

Regenerative agriculture is a set of land management practices that seeks to restore degraded soils with the goal of fighting global warming (discussed in Chapter 2), increasing soil biodiversity, and reducing water scarcity.

Unfortunately, many common practices are devastating to soil fungi. Clear-cutting the land for agriculture, timber harvest, or development disrupts microbial communities and alters the physical conditions of the forest floor. When farmers till their soils or apply fungicides, they harm the soil fungi and can reduce the activity of arbuscular mycorrhizae (see Chapter 3), which are important to the growth of crop plants. Over time, human activities such as these lead to erosion of topsoil, which is a major threat to human food security.

People who are encouraging a switch to regenerative agriculture include scientists, policymakers, farmers, and members of the food industry. Together, they encourage farmers to adopt practices that preserve the microbial communities in the soil, such as:

>> Use minimum tillage techniques or don't till at all

>> Incorporate compost to restore microbial communities in the soil

>> Manage grazing to minimize soil damage

>> Plant cover crops on fallow fields to prevent soil erosion

>> Rotate crops to prevent mineral depletion

Adopting these practices will restore a healthy fungal network to the planet's soils, helping to remove carbon dioxide from the atmosphere and improving our chances of continuing to feed a growing population.

Chapter **18**

Ten Resources for Learning More About Fungi

I f you love fungi, you aren't alone. People have foraged for fungi for thousands of years, sharing their knowledge of where to find edible mushrooms with their families and friends. College students studying biology have unexpectedly fallen in love with fungi and found themselves starting careers as mycologists. Whatever brought you through the door to the fungi party, other people are ready to make you welcome. In this chapter, you are introduced to several different ways you can expand your knowledge of fungi.

The North American Mycological Association

The North American Mycological Association (NAMA) is a nonprofit group that supports all things fungi through science, education, and environmental protection. Group members include all kinds of people with an interest in fungi, from professional mycologists to amateur foragers. The NAMA website (https://namyco.org) has many resources for people who want to know more about fungi, including:

» **Educational resources:** NAMA has materials to support education about fungi, whether to help you deepen your own understanding or share your love of fungi with children.

» **Links to local mushroom clubs:** Joining a mushroom club in your area is a great way to learn more about your local fungi (for more on this, see "Your Local Mushroom Club" that follows). NAMA has a map and a list with links to mushroom clubs in the United States and Canada.

» **Culinary information:** Under this topic, you'll find recipes, cookbook reviews, sources for purchasing mushrooms, and webinars on topics related to fungi as food.

» **Book reviews:** NAMA's book reviews can help you pick out the best books to support your interests.

Your Local Mushroom Club

Many communities have their own mushroom clubs or mycological societies, groups of people who meet to share their interest in fungi. If you are interested in foraging, joining your local club can be a great way to connect with people who have experience with the fungi in your area. Many clubs lead forays, or mushroom gathering trips, when the local fungi are ready to be harvested.

Clubs may also bring in speakers or sponsor events, such as an annual show. One way to see if your community has a club is to check the list of clubs affiliated with NAMA at https://namyco.org/clubs. Another way is to search the Internet for "your town" plus "mushroom club" or "mycological society."

The U.S. Centers for Disease Control and Prevention

If you want to know more about diseases caused by fungi, the CDC website has a subsection with well-organized information on this topic (www.cdc.gov/fungal/about/types-of-fungal-diseases.html). One of the nice features of the website is that, as you explore a topic, you can often choose whether you want information at the level for the general public or whether you want more in-depth coverage for health care providers.

FUNGIWOMAN

If you think fungi are beautiful and you want to brighten your days with gorgeous photos of fungi in the wild, then follow @fungiwoman on social media. Barbora Batokova combines her love of mushrooms and photography to create stunning and informative posts on the fungi she encounters in her local area (Pennsylvania) and during her travels. You can also visit her website at https://fungiwoman.com where you will find recipes, mushroom profiles, and information on her new book, *Hunting Mushrooms: How to Safely Identify, Forage and Cook Wild Fungi.*

Websites

Michael Wood, a past president of the Mycological Society of San Francisco, started MykoWeb (www.mykoweb.com) as a place to collect information on mycology and local mushrooms. The site has gorgeous photos and information on the mushrooms of California, including keys to help you identify them. Other useful sites include Michael Kuo's Mushroom Expert website (www.mushroomexpert.com) and Gary Emberger's Fungi Growing on Wood (archived at https://mushroomexpert.com/fungionwood). The late Tom Volk of the University of Wisconsin put together a very comprehensive website on fungi that is still accessible at https://botit.botany.wisc.edu/toms_fungi.

Podcasts

If you like to listen to podcasts while you're on the go, there are a couple of good ones that specialize in fungi-related topics:

>> The "Mushroom Hour Podcast" brings in experts to talk about an enormous variety of topics related to fungi, from the fungi of different regions to medicinal and culinary uses of fungi, the role of fungi in soil health, and ways we might use fungi in the future. You'll hear some great conversations, and you never know where the hour may take you next.

>> The "Mushroom Revival Podcast" focuses on the use of fungi to improve physical and psychological health and investigates the potential of fungi in technological innovations.

YouTube

YouTube has lots of great videos on specific fungi, foraging tips, and basic mycology. (Check out "Fungi: Death Becomes Them — CrashCourse Biology #39" for a literal crash course on basic fungal biology.) Or search "Anne Pringle fungi" to find an excellent three lecture series that introduces fungi and then then takes you through some of the research that her laboratory did on *Amanita phalloides* in the San Francisco Bay Area. For an amazing look at how fungi grow, try searching "slow motion mushroom growth" or "time-lapse decomposition." In addition to searching for specific things you're interested in, consider subscribing to some channels that regularly feature fungal content, like @PlanetFungi, @anna-identifies-mushrooms, @LearnYourLand, @mushroomwonderland1, or @SouthwestMushrooms.

Fungi-Friendly Books and Magazines

So many great books have been written about fungi that it's impossible to name them all. Instead, this list presents a few categories of useful books and suggests a couple of top picks in each category:

- **General mycology:** If you're interested in more details about the structures of specific groups of fungi or their life cycles, consider picking up a general reference book. *The Kingdom of Fungi* by Jens Henrik Petersen is a beautifully illustrated book that delves into the major groups of fungi and considers fungal ecology. *The Fifth Kingdom* by Bryce Kendrick is comprehensive but fairly technical. You can sample sections of the book and browse beautiful pictures on www.mycolog.com.

- **Field guides:** If you want to identify fungi either for personal interest or for foraging, you'll need a good field guide to the mushrooms of your area. You can get recommendations for which ones are best from a local mushroom club or on the NAMA website (https://namyco.org/publications/book-reviews). If you're in California, check out *California Mushrooms: The Comprehensive Identification Guide* (www.californiamushrooms.us). Other recommendations for regional field guides can be found on FungiKingdom.net at www.fungikingdom.net/field-guide-recommendations.html.

- **Cookbooks:** Mushrooms provide a delicious umami flavor and pack some powerful nutrition, so it's no surprise that people have written cookbooks that feature fungi as the main ingredient. *Wild Mushrooms: A Cookbook and Foraging Guide* by Kristen and Trent Blizzard has tips for foraging as well as recipes that feature wild mushrooms. *Shroom* by Chef Becky Selengut has recipes for delicious dishes that range from simple to complex, as well as great information on different types of mushrooms. *Cooking with Mushrooms: A Fungi Lover's Guide to the World's Most Versatile, Flavorful, Health-Boosting Ingredients* by Andrea Gentl has great tips for working with mushrooms and recipes to incorporate mushrooms into every meal.

- **Growing guides:** Paul Stamets wrote two of the classic guides on growing mushrooms: *Growing Gourmet and Medicinal Mushrooms* and *The Mushroom Cultivator: A Practical Guide to Growing Mushrooms at Home* (written with J.S. Chilton).

- **Magazines:** Britt Bunyard's online FUNGI magazine is a great read (www.fungimag.com), as are the two newsletters published by NAMA that you can read right on the website (namyco.org/publications): *The Mycophile* contains a wide variety of general-interest contributions from members, including articles, events, contests, recipes, and even a holiday gift guide. *McIlvainea* presents articles that take a more in-depth look at a wide variety of topics related to fungi.

iNaturalist

iNaturalist is a website and an app for your phone that lets you record your observations of the natural world and connect with the observations of other people. For example, you can take a picture of a fungus that you spotted and upload it along with the time you took the photo and the location, and the app will offer the choice to view suggestions of what it might be by comparing your image to others in its database. Other iNaturalist users can make comments on your post to help with your identification. The app is run by a nonprofit group with the goal of recording the biodiversity of the world around us and making that information available to scientists. The app is fun and free and has information on all forms of life. It's also an important tool in fungal conservation and community science.

Commercial Suppliers for Growing Fungi

Many commercial sites sell mushroom kits and other materials to help you grow fungi at home. I'm recommending a couple here that have good reputations. I also recommend looking for suppliers in your local area to save on shipping costs. Northspore (https://northspore.com) has a good reputation among the growing community and excellent educational resources. From the main page, click "Learn" to access lots of great information and helpful videos. The pink oyster mushroom kits from Forest Origins (https://forestorigins.com) and Hodgins Harvest (www.hodginsharvest.com) both did very well in a test by Consumer Reports in 2022. For more advanced growers who want to buy spores or mushroom cultures, Spore Works (https://sporeworks.com) also has a good reputation.

Index

H

About the Author

René Fester Kratz, Ph.D., teaches botany, microbiology, cellular biology, and global health at Everett Community College in Washington State. When not teaching, she enjoys getting out into nature in the beautiful Pacific Northwest, making arts and crafts of all sorts, and spending time with her family. Kratz is also the author of several other titles in the *For Dummies* series, including *Botany For Dummies, Molecular and Cell Biology For Dummies, Biology For Dummies,* and *Genetics For Dummies.*

Dedication

To my family who supported me through the writing of this book, and to the memory of my father, James William Fester, because he was a fun guy who loved a bad joke.

Author's Acknowledgments

Thanks to Matt Wagner of Fresh Books, Inc., for helping me find the opportunity to write this book. And thanks to all the great people at Wiley who made it happen.

A very special thanks to Barbora Batokova (@fungiwoman) for reviewing this book and for her inspirational photos that showcase the beauty of fungi.

Publisher's Acknowledgments

Acquisitions Editor: Alicia Sparrow

Senior Managing Editor:
Kristie Pyles

Development Editor:
Katharine Dvorak

Technical Editor: Barbora Batokova

Production Editor:
Magesh Elangovan

Cover Image: © NaturePL/
stock.adobe.com

12 301

 ⸱   ↲